PRACTICAL
FIELD ECOLOGY

Fig. 13.—Ecological map of part of a Marsh Meadow:
Pontfaen, Pembrokeshire, June 1940. Student Record by Miss
Erica Moore.

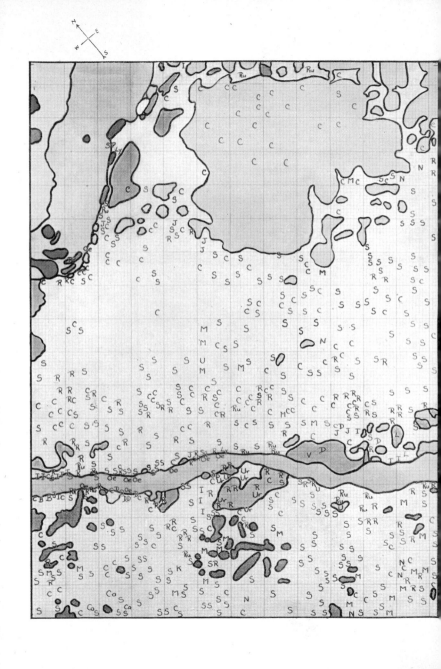

Community A. ▭   Pteridium ▭   Myriophyllum sp. ▭
                  aquilinum

uncus ▭   Ulex europaeus ▭   Crataegus ▭
  conglomeratus   monogyna

Open water ▭   Salix aurita ▭   Spiraea Ulmaria ▭

ris Pseudacorus ▭ Corylus Avellana ▭ Fraxinus elatior ▭

KEY TO COMMUNITY A. (FIG. 13).

Holcus lanatus   Rumex acetosa

Agrostis tenuis   Bellis perennis

Plantago lanceolata   Taraxicum vulgare

Plantago media
                              Veronica Chamaedrys
Lotus corniculatus
                              Achillea millifolio
Trifolium repens
                              Cerastium vulgatum
Prunella vulgaris

Sa. = Salix aurita   Ru. — Rubus fruticosus

Co. = Corylus Avellana   Sp. = Spiraea Ulmaria

Rf. = Ranunculus fluitans   Ca. = Catharinia sp.

J. = Juncus conglomeratus   V. = Valeriana sambucifolia

S. = Senecio Jacobea   Ur. = Urtica dioica

C. = Cnicus palustris   Oe. = Oenanthe crocata

M. = Circium lanceolatum   U. = Ulex Europaeus

R. = Rumex crispus   Ly. = Lychnis dioica

PRACTICAL FIELD ECOLOGY

A Guide
for the Botany Departments of
Universities, Colleges and Schools

By
R. C. McLEAN, M.A., D.Sc.
and
W. R. IVIMEY COOK, B.Sc., Ph.D.

LONDON
GEORGE ALLEN & UNWIN LTD
MUSEUM STREET

FIRST PUBLISHED IN 1946

PRINTED IN GREAT BRITAIN
in 11-Point Baskerville Type
BY UNWIN BROTHERS LIMITED
WOKING

PREFACE

THIS little book has been written as a guide to field survey work in connection with Plant Ecology. It also includes some simple work on soil and water analyses which might be carried out either in the field or subsequently in the laboratory. It is the outcome of many years' experience in conducting field classes and all the work herein described has been actually carried out in the field by successive generations of students under our supervision.

Field ecology is not difficult and is well within the scope of any student working for the Higher School Certificate, in fact such students usually have a great advantage in that they have more time to devote to this kind of work than the science student at a University. The necessary apparatus is not expensive and much can be constructed by the intelligent student in the workshop.

Very valuable information can be collected, especially where the same area is visited repeatedly over a period of years and such work is well within the compass of schools. Unfortunately in the past much ecological work lacked precision and field observations were often too vague to be of real value. It is hoped that this book will help to bring a more precise and systematic treatment of field ecology within the scope of students.

Nearly all the illustrations are original and the records and maps are all taken from actual surveys made by our students.

We wish, however, to thank Messrs. E. R. Watts of 108 Victoria Street, London, S.W.1, for the loan of the blocks of Figs. 2, 3 and 4, and Messrs. A. Gallenkamp Co., Ltd., of Sun Street, Finsbury Square, London, E.C.2, for the loan of the blocks of Figs. 34 and 35.

We wish also to thank Mrs. Ivimey Cook for the assistance she has given us in the preparation of the manuscript for publication.

In order to assist students in the identification of British Grasses when not in flower, an analytical key to the species of ecological importance, based on vegetative characters, has been added. A second key for the identification of the commoner genera of Marine Algae has also been added. Though not all the genera can be identified without section cutting, the recognition of most of the common genera ought to be possible in the field. Our thanks are due to the late Professor Harvey Gibson for permission to publish this key, which was constructed by him for the use of his own students, but was never published.

The larger Fungi are sometimes ecologically important as soil indicators, as mycorrhiza and in other ways. To assist in the recognition of the chief British genera, a third key has been prepared and is here reproduced. By its aid the generic names should be quickly discovered for use in the field, and specific identification, if required, may be left over for subsequent work in the laboratory.

No key for the identification of Mosses has been given. This is not because the group is ecologically unimportant, for Mosses are often very important in plant associations. Microscopic characters are, however, essential for their identification and this cannot be easily accomplished in the field. If, therefore, they have to be taken to the laboratory, it is better to consult a standard work, such as Dixon and Jameson's *Handbook of the British Mosses*.

The authors can speak from experience of the real keenness and interest which parties of students of average ability can show in field work which has been carefully planned along the lines here recommended. The change-over of duties between groups on successive days avoids boredom and stimulates competition, besides ensuring that all-round experience is enjoyed by everyone.

A most important feature of the plan is the prompt working up of the field records in the evening. Young people like jobs which are quickly finished and are usually ready and anxious to get on with their maps and plans.

The authors look back with nostalgic regret upon many evenings spent in hot and crowded sitting-rooms, with enthusiastic little groups occupying every available perch, amid a good-humoured babble of discussion and badinage, as each party strove to make its effort the best of the day.

Perhaps the best testimony to the value of field classes has been given by the numerous students who have confessed, "We never thought Botany could be half so interesting."

This book has been based on the supposition that the teacher is already acquainted with Ecology as a subject, understands something of its principles and appreciates its chief aims. It therefore confines itself to methodology and leaves untouched those broader questions which it is the province of a text-book to discuss. The problems which press upon a teacher are not all of them of a practical nature however. Before practical problems can even be thought of there must come the deeper considerations of the educational and scientific value of subjects, and of pedagogical limitations which are often the deciding factors in determining how far a study may be carried.

Those school teachers who would appreciate a discussion on these and kindred problems in connection with Ecology and who would like to have an exposition of its content and methods in a short and simple form cannot do better than to turn *Plant Ecology and the School* by A. G. Tansley and E. Price Evans. Professor Tansley's *Introduction to Plant Ecology* supplies a more ample exposition of the subject. Both these books are issued by the publishers of the present work.

University College
Cardiff.

March 1943

CONTENTS

LIST OF ILLUSTRATIONS

Chapter I

INTRODUCTION

THE analysis of vegetation by survey methods is becoming recognized more and more as a fundamental approach to field Botany. Gone are the days when students of Botany were conducted on a country ramble where a few of the most noticeable plants were collected at random and placed in a vasculum, to be identified subsequently and then pressed. While to a few ardent systematists this may have seemed the be-all and end-all of field Botany, by many it was regarded only as a rather dull way of wasting time, so that the party of students soon tended to dwindle into a small but faithful band of collectors, while the rest either amused themselves according to their tastes or sought the first opportunity of losing the party altogether. Such must have been the experience of many teachers of Botany from the great Linnæus downwards, with the result that field work became less and less popular.

Lack of interest in field work on the part of a class is largely attributable to the small proportion of Staff present in comparison with the students. Those immediately around the teacher will gain much information, but should a student stray away in search of any object which interests him, he will get no assistance or guidance in his work till he catches up with the party again. Teachers vary much in the way they conduct such classes; some go rapidly over the ground and think nothing of walking a dozen miles in the course of the day, thus leaving no one in the party time to wander about; others cover only a mile or so in a day and stop so frequently and for so long that the majority of the students get bored. Moreover it is an obvious fact that there is a limit to the amount of information which the average student can assimilate in a given time, especially when it is very difficult for him to write down the information. Even the identity of a dozen or so plants will be forgotten by most students by the time they reach home, so that instruction unduly prolonged is merely adding more water to an already saturated sponge.

In contrast to this type of field work is the ecological survey. Here a small, but instructive area is selected in advance with a view to demonstrating as many ecological principles as possible and the area is subjected to an intensive study. Every student in the party has a definite job to do, a job moreover for which he can have been trained previously in the laboratory. More than this, he is generaley working as one of a team and often in conjunction with another team doing another investigation. When the results are worked up in the evening it will be possible to check the accuracy of the work of one team against another and thus to encourage competition. It cannot be too strongly emphasized that the fostering of this competitive team spirit makes all the difference to the keenness of a class of students engaged on field survey work, as well as to the accuracy of the results produced.

The purpose of this book is to describe as fully as possible the way in which such an ecological survey can be carried out. It is assumed that the whole, or almost the whole, of the work will be done in the field, away from any laboratory, and consequently the methods described are essentially field methods. In some instances alternative methods might yield more accurate results, but, being unsuitable for field conditions, they have been omitted here. On the other hand the book does not deal with the purely theoretical side of plant ecology. This aspect of the subject has been treated by many authors. Elementary students will usually find a chapter devoted to the subject in their Botany text-book, and Professor A. G. Tansley's *Elementary Plant Ecology* provides a much fuller treatment, with hints on field work. University specialists need no direction to the standard text-books on British Vegetation, such as Professor Tansley's *The British Islands and Their Vegetation*. For teachers in schools who are considering the possibility of ecology as a school subject we recommend the book by Professor Tansley and Mr. Price Evans, to which reference has already been made in the Preface.

Experience over the last fifteen years in conducting ecological field classes among University students has enabled the authors to reject or modify many of the schemes which have been advised for survey work, in order to arrive at one suitable for class work without losing the necessary degree of accuracy. The methods and technique described in the following pages have been tried and proved effective by successive generations of students, and, while minor modifications are constantly being made, it may be said that the general scheme of work has stood the test of experience and has proved satisfactory.

Here at Cardiff it is customary to arrange each year, at the end of the Summer Term, an expedition to some suitable locality, lasting about a week, which is attended by all Final and Honours students. This means that approximately thirty students are involved and the organization of the survey work is based upon suitably employing that number. It is comparatively easy to employ a small group of half-a-dozen or so keen students, but, where a larger and more varied class is involved, the organization has to be far more elaborate. If therefore, in the following pages, certain points seem unnecessarily detailed it must be remembered that some simplification is possible where only a small group is concerned. On the other hand a small band of keen students will accomplish relatively more than a large class of mixed quality.

It must be pointed out that the survey of a particular site is finished and worked up in a day, so that during a week six or seven surveys are completed. The independent student must not think that it is impossible for him to get through the same amount of work because he is working alone. He will probably only attempt to analyse a single site, and it should be quite possible for him in the course of a session to obtain much interesting information.

The whole of the records and observations made during a survey are in themselves difficult to assimilate, except by those who have actually taken part in the work. In order, therefore, to make the work of wider value and useful for reference by subsequent students it is customary at Cardiff for a general report to be compiled by the member of the Staff in charge of the party, setting out certain general features of the locality selected for the survey work and collecting together and analysing the results obtained. In such a report it is also possible to criticize or comment upon any inaccuracies or mistakes not only in the work of the students but also in the equipment, so that, as far as possible, these may be corrected in the following year.

The whole of the results of each annual excursion are bound, the report and statistical matter in a small single volume, and the maps and large graph sheets in large folders. Since the number of large sheets is relatively few it is best to bind those of three or four years together to make a reasonably thick book.

In some instances photographic records of vegetation may be included in the report, but the production of really good ecological photographs calls for considerable skill and judgment. This can be readily appreciated from a study of many of the published photographs which, while appearing superficially to be quite nice pictures, tell nothing of the details of the vegetation. Few students are sufficiently experienced in photography to be able to do this work well. Moreover, a hand camera is usually valueless. Only a stand camera, with a lens well stopped down, combined with the use of appropriate light filters and consequently a long exposure, can give the desired results. Further, the long exposures require perfectly still weather when the vegetation is not in motion, a condition which seldom occurs. It, therefore, comes down to this, that good ecological pictures are of the greatest value but are very difficult to take, while second-rate photographs have little value as records and are only admissible as general indications to supplement proper surveys.

Chapter II

APPARATUS AND EQUIPMENT

BEFORE any serious survey work can be attempted a considerable amount of apparatus is necessary, but, while certain pieces must of necessity be bought, many can be made by the teacher or student while others can be improvised for the particular occasion.

One of the great problems which has to be settled at the outset is the method of transport of the apparatus from the laboratory to the survey site. The sharing out of the apparatus piece by piece among the students present is unsatisfactory for, however much care is taken, some essential piece is almost certain to be left behind, while the policy of packing it all into the teacher's private car, even if he is willing, is attended by the same objection.

The ideal condition is one in which on arrival at a new site, all the necessary apparatus is immediately available to carry out a complete survey, whatever the local conditions may be. In the Department of Botany at Cardiff this has been satisfactorily solved.

Six wooden boxes have been constructed and fitted to contain the apparatus necessary for carrying out all the experimental and survey work described in this book, with the exception of a few large pieces, such as the pickets and staves, which must be taken separately. These boxes have travelled many miles, in railway vans, on board ships, on the backs of cars and even in farm carts, without serious harm or damage to the enclosed apparatus. They contain all the apparatus needed for a class of thirty students, yet on the other hand they are reasonably portable and have, at times, been carried up mountain sides or to other difficult sites by the students themselves.

The idea of carrying ecological apparatus into the field by means of fitted boxes was first brought to the authors' notice by the late Dr. Rose Bracher, who used one for her ecological class at the University of Bristol. By subsequent elaboration of this idea the six boxes used at Cardiff came into being. All the construction work was done in the departmental workshop. It must be admitted that a professional carpenter could probably save weight by

FIG. 1.—Ecology Field Box A. Showing the apparatus it contains. For details see text.

using thinner wood for the internal divisions, though whether this would be an advantage is a moot point, for, though the box would be lighter, it would be less sturdy. It may well be that it is just this high degree of rigidity which has enabled the boxes to stand up to the rough handling and save the apparatus being damaged in transit.

Furthermore, since each piece of apparatus has its own compartment it is a matter of comparative simplicity to check over the boxes at the end of a day's work to see that all the apparatus has actually been returned.

It is now proposed to indicate briefly the contents of the boxes and the purposes for which the apparatus is used, detailed description being given in the appropriate chapters. It will be appreciated that as far as possible the apparatus has been grouped in the boxes for special lines of work. Thus Box D provides apparatus for physiological observations, and Box E for limnological work, so that neither of these boxes need be taken unless this type of work is contemplated. If no analysis of the soil is proposed, a policy to be deprecated, Box C need not be taken. It follows therefore that all the essential apparatus for a simple grid survey is contained in Boxes A and B.

Box A (Fig. 1), which measures 37 ins. × 14 ins. × 5ins. contains the following apparatus arranged as will be seen in the illustration.

4	66 ft. Surveyors' tapes	2	Levelling telescopes
1	Surveyors' chain	4	6-in. Valence squares
12	3-ft. wooden rules	2	150-ft. lines
4	3-ft. spring tapes	2	Soil trowels
1	Cross staff head	1	B.D.H. "4–11" Capillator set[1]
1	Clinometer compass	1	Bottle of distilled water
12	6-in. wooden pegs	1	Bottle B.D.H. "4–11" Indicator[1]
12	Iron spikes	2	Quadrat strings on winders

The tapes are made of linen and can be bought with a metal thread running through them which, while valuable for accurate work, because it reduces shrinkage, is unnecessary for ordinary work and increases the cost considerably. Care must be taken to see that tapes are carefully rolled up. The tape should be run between the two fingers of the left hand, while holding the case with the same hand, to ensure that no twists are rolled in

[1] Made by British Drug Houses Ltd., 1 Graham Street, City Road, London, N.1.

and also to keep a tension on the tape. Tapes must not be put away wet. A tape rolled up wet will cause the inside packing of the case to swell, with the result that, once in, no power on earth will get it out again, and it can only be freed by cutting the stitching of the case, thus ruining it for future use. If carefully used a tape will last for years, but in the hands of a careless worker it can be quickly damaged. There can hardly be too many tapes on a survey and at least four are desirable.

A surveyor's chain, divided into 120 links, is not essential, in fact it is very often removed from the box before starting on an expedition. It is a heavy lump of a thing, increasing the weight of the box considerably. A good chain weighs about 11 lbs., and for most purposes a tape answers equally well. In woods, among dense undergrowth, a chain can be more easily dragged than a tape and is more accurate, but in all probability the error of a tape compared with that of a chain is small when set against the personal errors of the individual students.

A dozen yard sticks are needed. These again may be bought, but quite efficient substitutes can be made from pieces of window-stopping laths which are about one inch wide and $\frac{1}{4}$ in. thick. One side is divided into inches and half-inches by lines carefully ruled in Indian ink, after which the whole lath is thoroughly varnished.

The spikes are surveyors' iron pins and are simply giant skewers. They can be made out of suitable lengths of iron wire about $\frac{3}{16}$ in. diameter. A loop is made at one end while the other is filed to a point. It is desirable to have a small length of brightly coloured stuff sewn to the ring as this makes it easier to find them again in thick vegetation. Eight should be taken.

Two trowels are included. These may be either those sold as fern spuds or common garden trowels. Experience has shown that the former, though smaller and more compact, suffer from the defect that sooner or later they bend and finally crack at the point where the upper rivet is inserted. Simple garden trowels, though more bulky, are on the whole more useful.

A number of small spring steel tapes will be found very useful for measuring small distances or objects. They are

FIG. 2.—Two types of instrument for laying out right angles on the ground. The brass Cross Staff Head (left) is fitted on to the top of a picket about 5 ft. high. The Optical Square (right) is held in the hand.

cheap to buy and take up very little room. Care must be taken to dry them thoroughly after use, for damp tapes soon rust and then will not close up properly.

The cross staff head (Fig. 2)[1] is a device for laying out right angles on the ground and is of great value in setting out a survey. It usually consists of an octagonal brass head, the alternate segments of which have a slit and a

---

[1] The optical square shown in Fig. 2 may be used instead of the cross staff head described above.

cross wire in them, so arranged that if in one segment the upper half of the slit is open and the lower half of the slit has a median cross wire in it, the opposite segment will have the same slit and cross wire but in the reverse position. Thus by viewing through any slit a cross wire running down the centre of the opposite slit will be seen. This octagonal head is provided with a short, conical, hollow support, which will fit over the top of a tapered pole of suitable length. The method of using the instrument will be described in Chapter III, p. 37.

Two sighting levels (Fig. 3) are taken, though with a small class one would be sufficient. These levels can be used to obtain very accurate measures of the slope of the

FIG. 3.—A useful and inexpensive form of Sighting Level.

ground. They are relatively cheap and overcome the need for a complicated levelling tripod and telescope. The level consists of a fixed-focus telescope and an enclosed mirror whereby the air bubble of a spirit level inserted in the top of the telescope can be seen through the telescope itself. As seen in the mirror the bubble moves up and down and a cross wire is inserted in the telescope in such a way that, when the image of the bubble is bisected by the cross wire, the level is horizontal. By looking through the telescope at a distant vertical scale, the instrument may be tilted until the cross-wire cuts across the bubble and the height on the vertical scale is read off at the same time. Further details of using this instrument will be given in Chapter IV, p. 48.

A compass is not an essential instrument though a very useful one. The type employed is a sighting clinometer compass (Fig. 4) with a brass strip hinged to fold over the top and provided with a viewing hole and a slit. In this way the compass may be sighted along a particular line and the angle between the line and the magnetic North easily read off. A clamping device is provided so that the swing of the needle can be slowed down, a point of some value, especially in windy weather.

By turning the sighting bar at right angles and using it as the base of the compass the instrument can also be used to determine a slope, making use of a second swinging

Fig. 4.—A Clinometer Compass. The instrument is shown in face view and also from the side with the sighting bar raised.

arm enclosed in the compass, the angle between this moving arm and the horizontal giving the angle of slope of the ground. This method can therefore be used as a rough check against the levelling observations, especially on steeply sloping ground, and also for estimating the height of trees, see p. 70.

The analysis of the vegetation within the area of the survey can be done best by selecting, either at random or in definite positions, small areas of ground in which a detailed examination of the vegetation is made. For this purpose squares of known size are most suitable. The four 6-in. valence squares are convenient (see p. 79), especially for analysing grassland or situations with a large variety of plants all of which are of small size. Where, owing to the larger size of the plants, bigger squares are necessary, areas may be pegged out by using 6-in. wooden pegs and joining them with a string to form a square. When the vegetation is very variable

FIG. 5.—Ecology Field Box B, which contains the grid strings.

in size, such as in a wood, larger areas still, up to 10 ft. square must be employed and the large grid strings are then utilized for marking out.

The equipment necessary for determining the hydrogen ion concentration, or pH, of the soil is also included. This consists of a bottle of the selected indicator, a bottle of distilled water, and a capillator set of standard colours by means of which the colour produced by the indicator in a soil solution may be compared with the colour of a standard buffer solution of known pH. For details of using the method see p. 94.

Box B (Fig. 5) contains the grid strings. The box is 28 ins. × 14 ins. × 6 ins. Sixteen of these strings are included, though in many instances an even larger number would be useful, and two more are usually substituted for the chain in Box A.

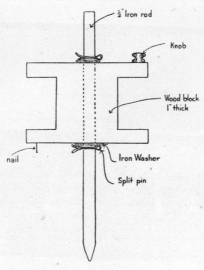

FIG. 6.—Diagram of a Grid String Winder, showing the method of construction.

Each string is about 150 ft. long and is made from good quality, fairly thick string, which is all the better for being waxed by drawing it slowly several times across a block of bees' wax. One end has a loop, which may be made either by tying or by attaching a brass ring. The latter is preferable. The other end is attached to a winder. The winders sold for garden lines are quite suitable, but an equally efficient and lighter substitute can be easily made at home. Take a piece of 1-in. board about 4 ins.

wide and cut out a rectangular segment at each end (Fig. 6). In the centre bore a $\frac{1}{2}$-in. hole running through the width of the wood. To one upper corner of the block screw a small knob and at the opposite lower corner fix a 1-in. nail projecting downwards. Cut off

FIG. 7.—Ecology Field Box C, which contains the soil-tins and soil-borers. One tin has been opened to show the waxed paper container.

about 12 ins. of $\frac{3}{8}$-in. iron rod and drill two holes in it, one about an inch from one end and the other $4\frac{1}{4}$ ins. from it. Into the upper hole fit a split pin, then place on the rod in order, an iron washer, the wooden winder, another washer and finally through the lower hole, a second split pin. Such a winder will be found quite efficient, especially if the whole has been well painted. It is advantageous to use bright colours, red or yellow, so that the winders can be easily found if left lying about on the site. These home-made winders are considerably lighter than the all-metal type sold for garden lines.

Care must be taken to see that the lines are thoroughly dry before being put away, otherwise, even if waxed, the string will soon rot.

Box C (Fig. 7) contains the equipment for taking soil samples for subsequent analysis in the laboratory. The box is 30 ins. × 14 ins. × 6 ins., and contains twenty-four tins. These are ordinary 4-oz. tobacco tins, which are particularly suitable for the work, especially if the ventilation hole in the bottom is soldered up. The tins may be painted inside and out to prevent rust. Each tin and its lid should be clearly numbered. The tins are numbered so that no error will be made as to the contents, while the lids are numbered to make it easier to pick out a required tin later. If, however, only the lid is numbered there is a risk that by an accidental substitution of lids the soil samples may be wrongly identified from the records.

The inside of each tin is lined with waxed paper. Newspaper is very suitable. Cut a number of pieces of newspaper to size and wax them by drawing them through a large shallow dish of melted paraffin wax. Select a tin or bottle which is slightly smaller than the soil tin and wrap a sheet of the waxed paper round it. Fold the lower end carefully in and leave the paper to harden. When cool remove the tin from inside and slip the paper tube into the soil tin. Then finally fold in the top of the paper carefully and put on the lid. A fresh sheet of waxed paper is used for each soil sample.

At the back of the box, that is at the bottom when it is being carried by the handle, lie the soil-borers. They are of two kinds. One is suitable for soft soil, the other for hard rocky soil. For soft soil, such as leaf mould, a 2-in. borer is used. It is made of iron and resembles a very large cork borer with a cutting edge and a ring at the top through which a wooden turning handle, or tommy bar, can be thrust. Soil samples up to 12 ins. below the

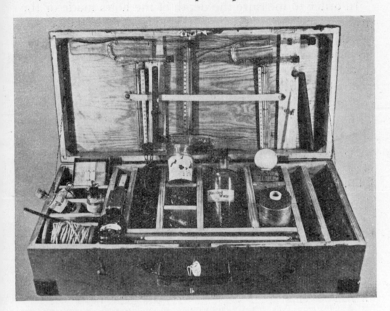

Fig. 8.—Ecology Field Box D, which contains the apparatus for field physiological experiments. For details see text.

ground can be taken with this, the distance being determined by the amount of the tube thrust into the ground. The tube is slotted on one side to aid the removal of the sample.

For hard rocky soils the auger borer is more suitable. This is simply a wood auger, 1 in. in diameter which is twisted into the soil with the aid of a wooden handle.

If the auger is pulled out vertically, without twisting, after reaching the required depth, a small sample of soil from the borer hole is brought up between the blades. A metal plate may be fitted around the auger so that as the latter turns the soil particles brought up fall on the plate and can be easily collected.

In order to measure the depth of the holes made or the distance the borer has been thrust in, a number of 2-ft. wooden rules are included in the compartment with the borers. These rules are also useful for making scale drawings of soil profiles or plant root-systems.

Box D (Fig. 8) measures 27 ins. × 13 ins. × 7 ins. and contains the various apparatus employed in the study of climatic and physiological factors in the survey area, as described in Chapter XI.

Its contents may vary somewhat according to the type of work contemplated. Normally the box contains the following apparatus:

1	Paper strip hygrometer	1	B D.H. "Soil Indicator" Capillator set
3	Atmometers		
3	Bottles of cobalt chloride paper	1	Light meter
		4	Stop watches
3	Boxes of standard colour paper on slides for matching cobalt paper	6	Student thermometers
		2	Delicate thermometers for Wet and Dry bulb
18	Bull-dog clips	2	Extra spring tapes
1	Bottle of 1,000 c.c. distilled water	1	Bottle of B.D.H. "Soil Indicator"

Each of these instruments is necessary for some particular experiment which can be carried out in the field. They relate mainly to atmometry, transpiration rates and a comparison of light intensities. In addition another Capillator set, with the appropriate pH indicator, is included as a reserve. The detailed methods employed in using this apparatus are given in Chapter XI, p. 94

and need not be repeated here. Where alternative experiments are contemplated other apparatus could be substituted and the contents of the box varied somewhat.

Box E (Fig. 9) is the largest of the Ecology boxes,

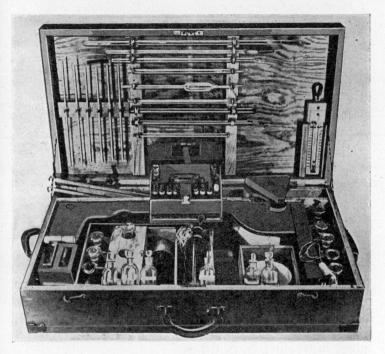

FIG. 9.—Ecology Field Box E, which contains the glassware used in limnological work. For details see text.

measuring 33 ins. × 19 ins. × 8 ins. It contains the special apparatus used in limnological work. This involves mainly the chemical analysis of the water and observations on physical factors affecting the water. Box E contains therefore, reagent bottles and chemical apparatus suitable for carrying out these tests in the field.

After some experience of this work the following apparatus has been found necessary:

2 Nessler cylinders, 50 c.c.
2 Glass rods
1 Distilling flask and condensing tube
2 100 c.c. beakers
1 Spirit lamp
1 Tripod and gauze
2 Stoppered bottles
1 Weighted bottle for water sampling
2 10 c.c. pipettes
2 Graduated 1 c.c. pipettes
2 100 c.c. stoppered measuring cylinders
2 25 c.c. burettes
1 B.D.H. "Universal "indicator capillator set
Plankton net with rod and tow rope
Grapnel with string
Six's maximum and minimum thermometer
Filter funnel and filter papers
Turbidity rod
Comparison Colorimeter

200 c.c. bottles of:
  N/50 Sodium hydroxide
  N/25 Hydrochloric acid
  N/100 Sodium thiosulphate
20 c.c. bottles of:
  Nessler fluid
  Phenolphthalein
  Methyl Orange
  Manganous sulphate
  Sodium hydroxide in Potassium iodide
  Sulphuric acid (conc.)
  Starch solution
Soloid tablets of:
  Ammonium chloride
  Zinc dust
  Sodium acid sulphate
  Potassium nitrate
  Potassium chromate
  Potassium permanganate
  Silver nitrate
  Soap
  Metaphenylenediamine sulphate

The detailed description of the use of this apparatus is given in Chapter X, p. 103.

Box F (Fig. 10) is the smallest, measuring 16 ins. × 13 ins. × 8 ins. It contains two microscopes in the two side compartments, while the centre division is used in various ways. For work in the evenings, when away from home, an electric lamp of suitable voltage is taken, if electricity is available; if not, a small paraffin lamp may be substituted. Slides, cover-glasses, dissecting needles, forceps and a razor are also included, together with a bottle of

Lactophenol and Cotton Blue. Sometimes a couple of dissecting microscopes are added, especially where floristic work is anticipated. When Box F is taken into the

FIG. 10.—Ecology Field Box F, which contains two microscopes, lamp and instruments for microscopical work. Alternatively, the central compartment may be used for books.

field in connection with limnological work, the lamp and reagents are removed and the space is filled with reference books for use in identification of the plankton. Two Rafter cells are also taken. These are designed to hold

C

exactly 1 c.c. of water in a film 1 mm. deep. By this means a quantitative estimate of the plankton population in the water sample may be obtained.

In addition to apparatus which can be carried in boxes, there are certain bulky articles which must be taken separately. These consist of a dozen 6-ft. surveyors' pickets, painted black, white and red in successive feet. In addition about 50 four-foot stout bamboo canes are needed. If these are painted red it will be easier to see them in dense vegetation. Both the pickets and the canes should be tied up in bundles of convenient size with leather straps. Finally one, or preferably two, levelling staves must be taken. These can either be bought or can be made by firmly glueing paper scales, marked in feet and inches, on to 2 in. by 1 in. battens, hinged in the middle to fold up. These paper scales are best bought[1], because the figuring has to be boldly printed if it is to be visible at a distance. One stave about 14 ft. and the other 9 ft. long are convenient, the former for hilly situations, the latter for working among undergrowth in woods.

[1] Supplied by Messrs. E. R. Watts (see p. 7).

Chapter III

SURVEY METHODS: THE GRID[1]

THE site of the survey having been approximately determined, four pickets are set out to form a rough square, indicating the corners of the area to be surveyed. It is desirable that, after this has been done, all members of the party should keep outside the area until their particular work has been assigned to them, in order to minimize damage to the vegetation from the trampling down of the plants.

A base line must now be set out. This may either be one side of the area or may pass through the centre. It is important that the base line should be visible from all parts of the area, consequently if there is much variation in the level of the ground it is best to set out the base line along a ridge. See Fig. 11.

Place a 6-ft. picket at one end, which may be either one of the corners already marked or some suitable point on a straight line between two corners. Then, with the aid of a tape, measure off a distance of 150 to 200 ft. Longer base lines are rarely needed, for they embrace too large an area to be analysed conveniently, though sometimes the base line may be considerably shorter. Place another picket to mark the end of the base line. It may now become necessary to modify the size of the area originally marked to make it correspond with the length of the base line. This is a valuable corrective to the tendency to select areas which are much too large.

With the aid of bamboo canes divide the base line into a number of equal distances, determined according to the nature of the vegetation. In a wood with much under-

[1] For the method of making a Plane Table survey, see Chapter XII, p. 156.

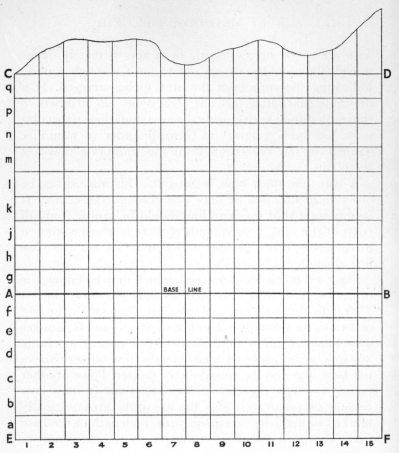

Fig. 11.—Diagram to show the method of gridding an area with strings. Each square equals 10 ft. by 10 ft. The Base Line is on the highest ground, across AB, and the margin CD has been extended to include an irregular boundary line. The distances are measured from CD.

growth 30 ft. may be suitable, while in grassland 5 ft. may be more convenient, because of the small size of the plants and the greater detail which will have to be put into the map. In general it may be said that the more uniform the vegetation the wider apart may be the canes. At this stage it may be found desirable to modify the length of the base line slightly to make it an even multiple of the intervals chosen.

It is now necessary to set out perpendiculars from the base line. This is most conveniently done with a cross staff head. Set up the cross staff on its support in place of the picket at one end of the base line and sight along to the picket at the other end, twisting the cross head on its holder until the cross wire is accurately in line with the picket. Now examine, through the slit at right angles to this, one of the corner pickets standing approximately at right angles to the base line and have it moved by an assistant until it comes into view on the cross wire, taking care not to move the cross staff during the process. Insert the picket when it stands directly in line as seen on the cross wire. This picket then should be at right angles to the corner of the base line. Check this by viewing again the distant picket on the base line to see that it is still in line with the cross wire. Remove the cross staff and its support and replace the end picket in the same hole.

If a cross-staff head is not available, a right angle may be obtained in the following way, using two tapes. Measure out along the base line a distance of 9 ft. and put in a cane. Slip the loop of one tape over this cane and loop the other over the corner picket. Measure off 12 ft. along the latter tape and 15 ft. along the former. Bring these two points together and insert a cane. This cane should be at right angles to the base line, in line with the corner picket. See Fig. 12.

Repeat the process at the other end of the base line and insert a corresponding picket. Four points will now have been set out forming a rectangle. If the base line runs

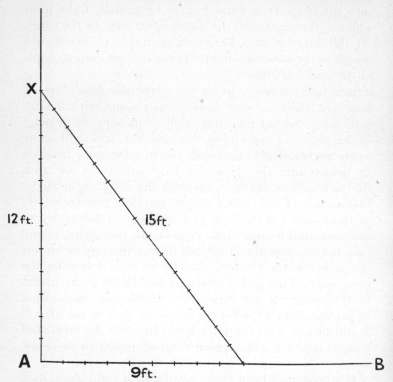

Fig. 12.—Method of obtaining a right angle to a Base Line AB. Two measuring tapes, 9 ft. apart, are run out to 12 ft. and 15 ft. respectively. They intersect at a point X. The line AX is at right angles to AB.

through the centre of the site it will now be necessary to sight two additional pickets on the other side of the line. This will become clear by reference to Fig. 11. Picket E is sighted in line with pickets A and C, while picket F is arranged in line with B and D.

To check these results measure the distances CD and EF, which should be each of them equal to the length of the base line AB.

Now measure off distances along each of these perpendiculars such that the whole length of each perpendicular is not more than 150 ft. and insert fresh pickets. These last pickets replace the original corner pickets which should now be withdrawn. A perfectly rectangular area is now enclosed by pickets ABCD or CDEF according to where the base line lies. In the latter case there are two additional pickets at A and B. Remember that all pickets not in use should be stuck into the ground at such an angle that they will not be mistaken for marking pickets. Never leave pickets lying on the ground, it is difficult to find them later.

It is practically essential that the base line should be parallel to two of the sides of the rectangle which is to be surveyed. If it was necessary, for topographical reasons, to choose a base line which was not parallel to one of the sides of the area originally marked out, this squaring up operation will result in turning the whole area through an angle, but this may be inevitable.

Starting from the base line now measure off distances along the remaining sides of the rectangle, equal to those into which the base line has been divided, and insert a bamboo cane at each point.

Next run out strings from cane to cane across the area, thus dividing it up into a series of squares, forming what is termed a *Grid*. It will be found most convenient if all the strings in the same direction are run out together. If there are not enough strings available to cover the whole area it is best to complete, if possible, all the strings one way, leaving only a few strings crossing them, since these can be moved on repeatedly as the work progresses.

It is important to make sure that the strings are drawn

taut and are well pegged down, and also that all parallel strings cross either over or under all those running at right angles to them. If they are mixed it is much more difficult to take up a single string later if required.

In woodlands, where it is impossible to set out strings because of trees and undergrowth, canes can be placed where the intersections of the strings should be, by sighting along the two lines at right angles.

The grid having been once set out the rest of the work on the survey can begin, and the mapping, levelling, belt and profile charts can be easily located by observing the squares in which they have been done.

It is sometimes desirable to map out an irregular area. In this case the same grid system is adopted, the irregularities in outline being measured back to the nearest grid square, as is shown in Fig. 11.

Chapter IV

MAPPING

EVERY student should be provided with a field note-book. This should be of the reporter's form, hinged at the top and about 6 ins. by 4 ins. in size. It should contain alternating sheets of one-inch squared and of plain paper. A rubber band should secure it and it should be provided with an attached pencil and with a length of string which can be looped over the neck, so as to leave the hands free. If students are numerous it is worth while getting a supply of such note-books made up by a firm of printers and bookbinders to one's own specification. If half a gross can be ordered at one time they will cost no more than a purchased book.

The mapping of the vegetation covered by a grid is a perfectly straightforward process. Using the field note-book make a note of the lengths of all the sides and of any other important measurements, such as the distance of an irregular outline from a grid string. Then construct a rough plan of the area making each large square of the squared paper equal to four squares of the grid and number each square as shown in Fig. 11.

Next go over the area, carefully noting the main characters of the vegetation, the different plant communities included and any other objects of special interest. The compass bearing of the base line should also be taken at this stage.

The mapping party may now split up into separate groups of one or two students. One group may start with the series of squares a1 to a15 and go on to b1, another may start with c1 and go on to d1 and another start with e1 and so on.

Take a fresh sheet of the note book and allow a page

of graph paper for each square of the grid. Write at the top of the page the number of the square and then draw in pencil the area of the square to a scale previously agreed among all the members of the group, so that each square of the grid will be mapped on the same scale.

Examine the vegetation of the square, carefully measuring the distance of any particular feature from the two nearest strings at right angles and plot that point on your graph paper. In this way the position of all trees and larger shrubs can be located, while the boundary of a plant community can be represented by a line passing across the square at the correct proportional distance. Record on the graph paper the names of the dominant plants in each community mapped and the name of each tree or shrub which is independently marked. If this becomes confusing use numbers instead and write a key on the opposite plain page of the note-book.

Now pass on to the next square and make another map on the next sheet of graph paper in the note-book, and so on till all the squares of the grid have been mapped.

Of course any special feature, such as a ditch or a path, should be carefully drawn in its appropriate position and size. Exposed rocks or bare ground should also be carefully marked and distinguished by selected symbols. In woods it is desirable to indicate, where possible, the area of shade cast by the trees, since this may have a marked influence on the ground vegetation.

The field maps having been completed, and after a check by the leader to see that all the squares of the grid have been mapped, the remainder of the work can be left till the evening in the work-room, when the final map is constructed.

Large sheets of graph paper, about 18 ins. by 24 ins., will be found convenient on which to reproduce the maps (see Fig. 13), though if great detail is desired a larger size of paper can be employed. A scale as large as possible should be arranged on the paper to fit the area of the survey. Thus, if the area is 150 ft. square and the size of the paper is that mentioned above, a 1-in. square of the graph paper may be made to equal a 10-ft. square of the map, that is, usually, one grid square.

The recorded map of each square is then carefully transferred to the appropriate position on the map, and, if the work has been done accurately, each zone marked in one square should join up exactly with the corresponding zone on the adjacent squares. Similarly a path or a stream should follow a natural course as it passes from one square to another.

After the outlines of all the plant communities have been mapped in this way and all outstanding features such as trees have been plotted in, the outlines should be drawn in with waterproof Indian ink. The sides of the area and the base line should be ruled in and the scale clearly marked along the side of the area. If the outline of the area is irregular it must be carefully drawn in by calculation from measurements made in the field.

Symbols may sometimes be employed to indicate special plants or plant communities. These may be arbitrary, but a key to any such symbols must be drawn at the side of the map with the Latin names of the plants to which they apply.

Next the position of the profile and belt transects must be indicated by ruled lines, and the position of quadrats marked. Random analyses, however, cannot be recorded on the map.

The map must next be coloured. The aniline dyes used for colouring lantern slides will be found quite suitable if they are sufficiently diluted with water. Bright colours are best avoided, though contrast is desirable. The paint should be put on as a wash and some skill is needed not to get it patchy. Each separate plant community should be represented by a different colour. Trees and the area of their shade are best marked in Indian ink before the paint is applied.

A key to the colours is included somewhere in the margin. Small squares or rectangles are ruled out in Indian ink one below the other. They may be made about two small squares high and three or four squares long. Each colour must be provided with a separate square. In line with the squares the names of the dominant plant of the community together with any of the co-dominants, are then written in Indian ink, and sufficient species must be included to indicate clearly the differences upon which the separation of plant communities has been based.

Finally the name of the district and of the site are printed at the top of the sheet together with the compass bearing. The group may then sign the map in the right hand bottom corner and record the date when the survey was made.

If it is intended to revisit the site at a later date it may be desirable to leave permanent corner posts driven into the ground. These should be made of 2-in. × 2-in. square pegs, preferably of oak, which have been well creosoted. Their length will depend on the locality. If the site is on private ground where there is little likelihood of the pegs being interfered with, they may be about 3 ft. long with about a foot driven into the ground. If, on the other hand, it is on public ground, the only chance of preserving the pegs is to drive them in flush with the soil and they should not be more than 18 ins. long unless the ground is marshy.

Great difficulty is often experienced in rediscovering the position of these pegs, which become buried under soil and vegetation. To help to recover them the tops

should either be marked or provided with an attached
lead disc bearing a letter corresponding to letters marked
on the map. Bearings should also be taken on suitable
land-marks or else distances accurately measured from
trees or other permanent features of the area. These data
should be accurately recorded on the map.

Chapter V

LEVELLING

IF full information is to be obtained to indicate the relation between the slope of the ground and the alterations in the composition of the vegetation, great accuracy in levelling the site may be necessary. On a salt marsh, for example, very slight differences in level will determine whether a particular area dries quickly or whether it becomes a drainage channel for land which may be only one inch higher. On the other hand, in a wood, differences of a few inches will probably produce little or no variation in the vegetation.

It follows therefore that due consideration must be given to these factors in deciding the method of levelling employed. At the outset it is usually best to level along the base line, starting from what appears to be the lower end. After that, all points of intersection of the grid strings are also levelled and calculated in relation to the initial point on the base line, which may be regarded as the *Zero Point*.

This zero point frequently has to be purely arbitrary, for in simple survey work it is rarely possible to relate it to any fixed datum point on the Ordnance Survey. Over a small area, however, it is relative rather than absolute levels which are important in ecology and it usually matters little what the actual altitude in feet and inches really is. Maritime and alpine areas are the chief exceptions to this. In maritime localities it is occasionally possible to take a level from some mark showing the mean low-tide level. In the latter case it is sufficient to get an aneroid reading of the mean altitude of the area or to calculate it from the Ordnance Survey.

There are various instruments devised for levelling

work. An elaborate tripod level may be desirable for very accurate work, but it is neither suitable nor necessary for student surveys. For such work the simple telescopic level described on p. 23 will be found quite satisfactory. Such a level may be held against a picket at a known

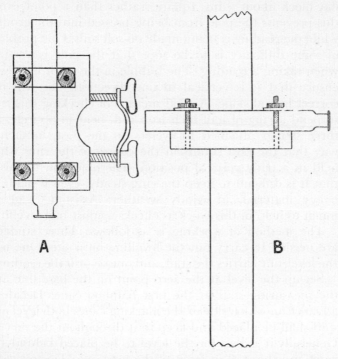

FIG. 14.—Clamping Device for Levels. A, horizontal view, B, vertical view.

height above the ground and used with satisfactory results, but it is more convenient and more accurate to construct the simple fitting shown in Fig. 14. This consists of a clip which can be clamped with screws to an upright pole which has been marked in feet and inches. It carries a cradle in which the level fits tightly. This cradle should be so constructed that its base is at right angles to the upright post. The fitting should be adjustable as to height, in order that it can be moved to accommodate workers of different stature.

It is preferable to use a post whose base is shod with a flat block about 3 ins. square, rather than a point, since this prevents the pole from being pressed into the ground while the reading is being made on soft soils. One problem of some difficulty is to be sure that the pole is vertical when taking a reading. The bubble in the spirit level will ensure that it is vertical in one direction, but will not correct lateral slope. A good method of checking this is to suspend a ring of half-inch iron rod, bent in a circle, by three strings attached to eyelets on the post, in such a way that the post stands in the centre of the ring when held in a truly vertical position. The objection to this is that it is difficult to keep the ring steady, even if made of heavy material, in windy weather. Assistants can do much to help in this work as checks against bad levelling.

The method of working is as follows. Three students are needed to carry out the levelling on a site. One uses the level, one carries the staff, and one records the readings.

Set up the level at the zero point on the base line and the measuring staff at the first bamboo cane. Decide in advance on which side of the marking canes both level and staff shall be placed and keep to it throughout the survey. Generally it is best for the level to be placed behind the cane and the staff in front of the next one. This prevents the cane from obscuring the scale on the staff, while, the level being higher than the adjacent cane, the latter does not get in the way. Set the level at a fixed distance up the pole, say 4 or 5 ft., and clamp it firmly. Place the pole in position and move it until the bubble in the level cuts across the horizontal cross wire (see Fig. 15). The student holding the staff can check that the level is being

held laterally vertical. Meanwhile the recorder can check
that the staff is also being held vertically by standing
back at right angles to the line of sight. That the staff is

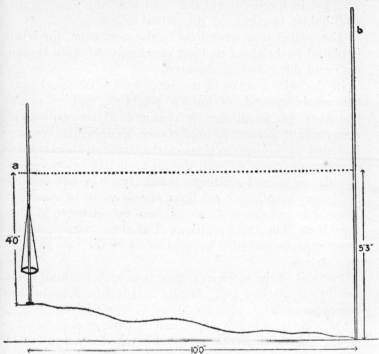

FIG. 15.—Method of using the Sighting Level. The level (a) is fixed at 4 ft.
vertically from the ground. The measuring staff (b) is held vertically at a
distance of 10 ft. The observed reading on the scale of the staff is shown as
5 ft. 3 in., indicating a drop of 1 ft. 3 in. The ring shown suspended from the
level picket is to aid in placing the staff truly vertical.

being held vertically in a direction at right angles to
this is checked by leveller and staff holder respectively.

Looking through the level, with the bubble cutting the
cross wire, the height on the scale on the staff should be
easily read off, but if the distance between the level and

D

the staff is considerable there may be difficulty in reading the actual figures. If the student holding the staff has a sheet of white paper and slowly raises it until it is seen by the leveller to cut the cross wire, it is easy for the staff-holder to check off the actual height.

The staff is now moved on to the next cane, the level is placed in the exact position previously occupied by the staff and the process is repeated.

In this way a series of measurements is obtained over the whole ground, at known positions, and in many instances this is all that is required. If the ground is uneven or if features of importance occur which are not included at the intersections of the strings—for example, if a stream passes through the area—it will be necessary to take additional readings. Similarly on a salt marsh, as already mentioned, readings may have to be taken at more frequent intervals to explain the changes in the vegetation. The exact positions of all these extra observations must be recorded by measuring to the two nearest grid strings.

The rest of the work in connection with levelling consists of calculations and plotting and is best done in the work-room later.

On a sheet of graph paper rule out the grid in pencil, on the same scale as that adopted for the map, and mark with a dot all points where level-readings were made. Now calculate the heights at all these points in the following way (see Fig. 16). Suppose the staff reading at the first cane along the base line was 4 ft. $2\frac{1}{2}$ ins. when the levelling telescope was fixed at 4 ft. This will mean that between these two points there has been an actual drop in the ground of $2\frac{1}{2}$ ins. Mark this point on the graph paper as $-2\frac{1}{2}$ ins. When staff and level were moved on to the next cane the staff reading was 4 ft. 5 ins. This means that at the second point the actual level was

4 ft. 5 ins. — 4 ft, a drop of 5 ins. in relation to the last point or — $7\frac{1}{2}$ ins. in relation to the zero point. The

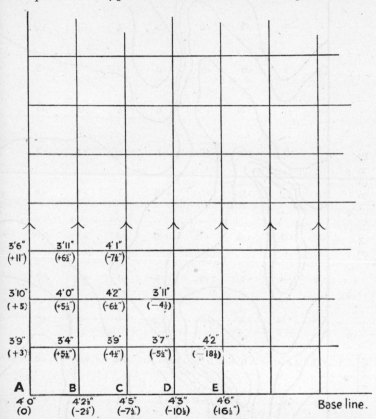

FIG. 16.—Diagram showing the observed staff readings over part of the area of a grid survey, taken in the order A, B, C, D, etc., and in the direction indicated by the arrows. The numbers in brackets give the actual levels in relation to the zero point A.

same calculations apply to all points along the base line.

Now turn to the perpendiculars. The first, starting at the zero point, will be calculated in exactly the same

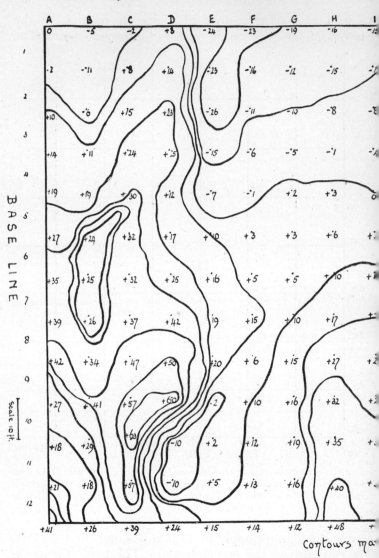

FIG. 17.—Contour Lines constructed on the basis of level readings
— showing a drop in level and + showing a rise. The c

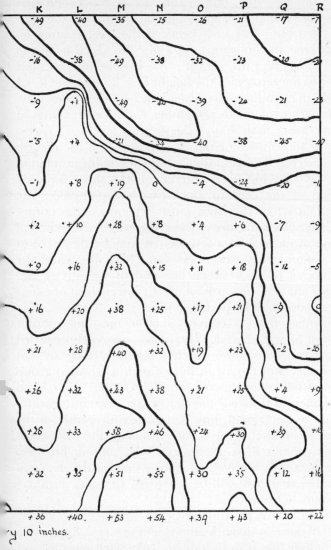

y 10 inches.

readings obtained at the points observed are marked in inches,
e obtained by joining points of calculated equal level.

way as the base line, but in the case of the second perpendicular remember that the initial position of the level
is not zero but $- 2\frac{1}{2}$ ins., hence if the level reading on
the staff was in this case 3 ft. 4 ins. the visual height of
this point is 3 ft. 4 ins. $-$ 4 ft., that is a rise of 8 ins., but
relative to the zero point the height is $+ 8 - 2\frac{1}{2}$, i.e. $+ 5\frac{1}{2}$
ins. All the readings are calculated in the same way.
The important point to remember is that all readings of
the scale lower than that of the sighting level itself represent *rises* and must be entered as positive on the record,
and vice versa. The actual value of each point in relation
to the zero having been entered on the sheet the construction of the contour lines can be attempted. The
distance apart of the lines must depend on the nature
of the site. On a flat area where the whole difference of
level is not more than a few feet, it may be desirable to
put in contour lines every 3 ins. On a hilly site, where the
difference of levels between one end of the survey and
the other may be upwards of 20 ft., intervals of one foot
or even more may be permissible.

Remember that to convey a clear meaning contour
lines must be well spaced. Where they are closely crowded
together and follow one another regularly they are not
all necessary and the mean line of such a crowded group
would be better and less confusing. On the other hand
if the lines are taken too far apart, certain features
obvious on the ground may be omitted altogether, if the
differences in level involved are less than the differences
between contour lines. If, for example, contour lines are
drawn in at 2 ft. intervals, it means that no difference of
level of less than 2 ft. will be recorded. In many instances
this would be disastrous.

The method of plotting the contour lines is perfectly
simple if done systematically (see Fig. 17). Suppose a
6-in. contour line is to be put in. Take each parallel and

perpendicular in turn and examine it. Supposing two adjacent points have heights of 3 ins. and 12 ins. respectively, it means that, assuming the slope to be uniform, there will be a point somewhere between these two which would have read 6 ins. That point will be three-ninths of the distance on the graph paper between the point marked 3 ins. and the point marked 12 ins. It is perfectly easy to estimate a point one-third along the line.

When the 6-in. contour has been put in, the other successive lines are plotted in the same way. The contour lines are obtained by joining together points of equal height, remembering that *no contour lines can ever cross one another*.

In this connection it should be remembered that contour lines may frequently describe closed areas and do not necessarily pass out at the margins of the map.

A sheet of tracing paper is now placed over the graph paper. The outline of the area, that is the outline corresponding to the map, is ruled in Indian ink, the contour lines carefully traced on it and the values marked in the margin. The position of the base line should be marked to enable the sheet to be orientated with the map. The name of the locality and the date should then be printed both on the tracing sheet and also the constructional graph sheet, after which both should be signed in the bottom right hand corner by those who did the work.

A check can now be applied by superimposing the tracing of the contour lines on the map. If the work has been done accurately the two should agree closely with one another. A pond, for example, marked on the map by a change in vegetation, should be clearly indicated on the levels tracing by a change in the contour lines which should more or less accurately follow the shape of the pond as shown in the map. Other similar checks can be applied, and if the work of both groups has been carefully done the levels tracing should explain some of the vegetation changes shown on the map. It must be remembered, however, that slope is not the only factor which influences the character of vegetation, so that we must not expect the contour lines and the changes in vegetation always to follow each other exactly. On the other hand conspicuous features should very closely follow the outline of the contour lines if the work has been well done.

Iris pseudacorus; Sium latifolium; Galium saxatile
Iris pseudacorus; Sium latifolium; Mentha aquatica
Iris pseudacorus; Sium latifolium; Galium saxatile
Iris pseudacorus; Sium latifolium; Epilobium palustre
Sium latifolium; Epilobium palustre; Veronica Beccabunga
Iris pseudacorus; Sium latifolium; Veronica Beccabunga
Iris pseudacorus; Sium latifolium; Veronica Beccabunga
Iris pseudacorus; Menthe aquatica; Epilobium palustre
Ranunculus acris; Mentha aquatica; Epilobium palustre
Sium latifolium; Galium saxatile; Mentha aquatica
Mentha aquatica; Epilobium palustre; Trifolium repens
Rumex longifolius; Trifolium repens; Mentha aquatica
Rumex longifolius; Mentha aquatica; Poa palustris
Myosotis scorpioides; Mentha aquatica; Poa palustris
Epilobium hirsutum; Rumex crispus; Potentilla Anserina
Polygonum Persicaria; Epilobium palustre; Milium effusum
Poa palustre; Milium effusum; Veronica Beccabunga
Veronica Beccabunga
Iris pseudacorus; Polygonum Hydropiper
Veronica Anagallis; Poa palustre; Veronica Beccabunga
Polygonum Hydropiper; Mentha aquatica; Epilobium palustre
Sparganium ramosum; Veronica Anagallis; Epilobium palustre
Sparganium ramosum; Veronica Anagallis; Mentha aquatica
Sparganium ramosum; Elodea canadensis; Callitriche palustre
Sparganium ramosum; Elodea canadensis; Nitella opaca
Sparganium ramosum; Elodea canadensis
Potamogeton crispus; Elodea canadensis
Potamogeton crispus; Elodea canadensis
Potamogeton crispus
Elodea canadensis
Elodea canadensis
Polygonum Hydropiper; Epilobium palustre
Polygonum Hydropiper; Mentha aquatica; Glyceria aquatica
Polygonum Persicaria; Epilobium hirsutum; Sium latifolium
Iris pseudacorus; Veronica Beccabunga; Sium latifolium
Urtica dioica
Urtica dioica
Urtica dioica
Iris pseudacorus; Myosotis scorpioides; Rumex crispus
Iris pseudacorus; Myosotis scorpioides; Polygonum Persicaria
Ranunculus repens; Plantago media; Cerastium vulgatum

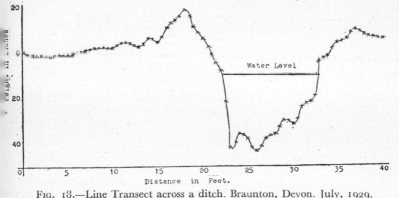

FIG. 18.—Line Transect across a ditch. Braunton, Devon. July, 1929.

facing page 56.

Chapter VI

TRANSECTS

TRANSECTS are designed to show graphically the changes in the composition of the vegetation along a selected line across the survey area. They are therefore definitely positional, in contrast to random sampling, and their situation must be carefully selected. If the site has been studied in advance it is customary to arrange for the perpendiculars to the base line to cut through successive communities as far as possible at right angles to their boundaries. If this has been previously arranged the transects may follow the line of the perpendiculars, and in practice this is generally done. The number of transects made on any one site must depend on the time available and the number of students who can be allocated to the work. Unless there is a great diversity of plant communities over the area one transect is generally sufficient.

This transect is best done along one side of the survey area, and the students should work from *outside* the area so that any trampling down of the vegetation will not damage the area and complicate the work of the mapping party or other groups.

Several kinds of transects have been devised to illustrate the character of the vegetation. The simplest is the *Line Transect* (see Fig. 18). This consists of a record of the plants occurring along the line, or grid string, forming one side of the survey area. In practice of course it is virtually impossible in a dense vegetation to make such a record complete, or if made, to reproduce it afterwards. Consequently some arbitrary selection must be made. The most general method is to record only those plants which occur at fixed distances along the line, say every 3 ins., 6 ins., or 12 ins. This is assisted by stretching a measuring tape parallel to the transect line. Such a method reduces the number of individuals recorded to one which can be later represented in the scale drawing. The main objection to this method is that it introduces a personal error which may become serious. Several plants may be found almost equally close to the point where the record is to be made. A selection is therefore necessary. Probably the most conspicuous plant will be recorded. Alternatively a plant recognizable at sight will be recorded in preference to one that is unknown; a low growing procumbent species will be overlooked and a tall adjacent one recorded. Thus, though the whole area may be frequented by a small, inconspicuous plant it may not appear in the list of species at all, simply by being overlooked.

In mapping a line transect, the levels along the transect are first plotted, using at least one small square horizontally to represent the distance between each point at which a plant record was made. Increase the vertical scale to several times that of the horizontal. Then write in, vertically, the name of the plant recorded at each point along the line. Should any special feature, such as a large rock, occur in the line, record both its position and, as far as possible, its size. Remember that the protection afforded by even a relatively small rock may be quite pronounced; in fact a micro-climate may be produced and the plants occurring in close proximity to the rock may be quite distinct from the rest of the surrounding flora.

The second, and better method therefore is to make a *Belt Transect*. For this method a strip rather than a line is used. Usually a yard strip is selected and a second string is run out, parallel to the side string and 3 ft. from it. With the aid of yard sticks the belt can now be divided into 3-ft. squares, each of which is examined separately.

Various methods can be employed for this examination. The older method was to regard each square as a quadrat and to plot-in the position of each plant in the area on squared paper in the field note-book. The method is straightforward and is an adaptation of the simple Quadrat, under which heading a description of the method will be given (see p. 74) (see also p. 64).

A more telling and instructive method is to work out the *Cover Index* for each square in the following way. Special ruled sheets are generally used (see Fig. 19), which can be either printed, or duplicated from a typewritten stencil, or they can be ruled up by the student as he goes along. In the horizontal column across the top is written the name of each species occurring in the square, while in the vertical column, on the left, is placed the number of the square measured off from the base line.

The percentage of the square occupied or covered by each species is then estimated by measurement and eye as accurately as possible. Such a method involves the identification of every species in the square, and this in itself is a very valuable training in the use of a Flora. The identifications must be accurate and the Latin names used; moreover, the species must be determined in every case, for in many instances the transition from one association to another involves the change from one species of a genus to another. A record of "Rush" or "Sedge," for example, means little or nothing. Rushes may occur all over the area, but the species may change as one passes from dryer to wetter ground. In fact in all ecological work the use of English names cannot be too strongly deprecated. It would greatly advance botanical studies if the English or common names of plants were completely excluded from scientific usage, however attractive they may be to the folklorist or the poet. Students should be taught once and for all the Latin

SWANAGE. DORSET. JUNE - JULY. 1938.

BELT TRANSECT.

Student: E.S.L. Jones.   Date: June 26th.

Sheet No.1.

Plant Community surveyed: Grass Downland, Ballard Down.

Serial No. of Quadrat	Poterium Sanguisorba	Dicranum scoparium	Asperula Cynanchica	Festuca ovina	Medicago lupulina	Plantago media	Cladonia pyxidata	Thymus Serpyllum	Lotus corniculatus	Viola hirta	Hieracium Pilosella	Galium verum	Carex glauca	Brachythecium purum				
1.	25	8	5	40	5	4	8	4	0	1	8	0	0	0				
2.	24	7	9	41	4	4	6	3	0	2	0	0	0	0				
3.	23	4	14	48	0	0	2	6	0	2	0	0	0	0				
4.	38	2	10	43	0	0	2	3	0	2	0	0	0	0				
5.	45	2	5	38	0	0	2	4	0	0	0	3	0	0				
6.	41	4	3	43	0	0	2	2	0	0	0	5	0	0				
7.	29	8	9	44	0	0	2	8	0	0	0	0	0	0				
8.	22	28	10	43	0	0	3	2	0	0	0	0	0	0				
9.	12	23	17	43	0	0	3	2	0	0	0	0	0	0				
10.	19	10	13	54	0	0	2	2	0	0	0	0	0	0				
11.	45	3	4	40	0	0	0	5	3	0	0	0	0	0				
12.	43	4	3	39	0	0	0	3	10	0	0	0	0	0				
13.	38	6	4	35	0	0	0	0	7	0	10	0	0	0				
14.	25	7	3	45	0	0	0	0	7	1	12	0	0	0				
15.	32	3	6	49	0	0	0	0	2	2	6	0	0	0				
16.	35	6	4	50	0	0	0	0	5	0	0	0	0	0				
17.	37	8	3	40	0	0	0	0	4	0	4	0	4	0				
18.	30	3	0	39	0	0	0	6	4	0	12	0	6	0				
19.	10	0	0	54	0	0	0	8	0	0	18	0	10	0				
20.	5	0	0	45	0	0	0	10	0	0	26	0	10	4				

Fig. 19.—A copy of the working sheet of a Belt Transect, showing the way in which the sheet is ruled. The area was a grass downland on Ballard Down, Swanage, Dorset. The names of the species have been typewritten for the sake of clarity. The numbers in the first column give the number of each square yard, counting from the Base Line. The numbers in the subsequent squares give the estimated percentage of each square yard covered by the particular species.

names, for the English name embraces too often a complex of species differing from one another just as much as plants with different English names.

It will be seen that to achieve a percentage estimation of cover it is necessary to regard the area in single layers. There may be several layers of plants present, but each must be treated separately. For example, bracken frequently forms a separate layer above a ground flora, and may cover 80–100 per cent of the area. Such a layer cannot be included in the composition of the ground flora, which would then obviously have to be represented by something more than 100 per cent. In a wood with trees, shrubs and ground flora, there may be several such strata to consider.

If the examination of the belt transect is to be made complete, a reading of the hydrogen-ion concentration of the soil at 2 ins. and 6 ins. below the surface should also be taken, the soil sample being collected from the centre of each square. See p. 94 for a description of the method used.

The results of such an investigation of a belt transect should be drawn up on graph paper, ruled as usual in inches and tenths. A separate histogram is made for each species, the graphs being arranged under one another. This is accomplished as follows. Let one small square vertically represent a percentage of 10. Then each graph will be only one large square high and can be separated from the next by a space of one large square. Rule a line down the sheet at a sufficient distance from the left-hand margin to allow the names of the species to be written in.

On the horizontal scale allow one small square to equal one 3 ft. square of the belt. Plot each result as a block graph, filling in the percentage found in each square as a solid block with Indian ink on the base that represents that square. Repeat for every species found along the transect (see Fig. 20).

Below the last species a graph should also be plotted showing the levels above and below the zero point, using a fairly large vertical scale, though not too large in relation to distance. Finally, at the bottom, the values of the pH measurements at 2 ins. and 6 ins. depth should be plotted-in using different coloured inks.

Much useful information as to the effects of slope and of pH upon the vegetation may be obtained by the study of such a chart if it has been accurately and carefully done. In vertical terms, the total area blacked-in for each square of the transect should be the same, i.e. 100 per cent, but it will be surprising to find how this is made up and how it varies from one part of the belt to another. One species may die out only to be replaced by another. Plants which are dominant at one end may be absent entirely at the other and so on. Two or more species may be found which appear to keep together, their occurrences and disappearances running parallel, thus suggesting that they are influenced by the same set of external factors. In other cases the distribution may be reciprocal, one disappearing as the other comes in. Such observations bring to light most interesting problems for further study.

If several layers have been distinguished, they must be graphed separately. Thus bracken may occupy only a small area of the ground vegetation, being represented only by its petioles, but in the upper layer, in the region of the fronds, it may cover possibly 75 per cent of the area, only allowing 25 per cent of the ground direct access to the light. These percentages must then be recorded separately either above or below the graphs of the ground flora, but in the same vertical line; though it is better to block them in with a different colour. The tree or shrub

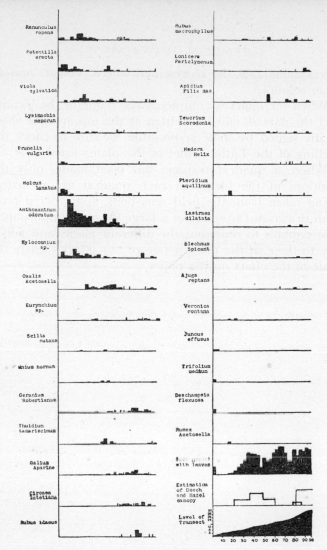

FIG. 20.—A Belt Transect Histogram showing the results of a survey across a sample of woodland in Llanerch Wood, near Newport, Pembrokeshire. Each small square of the graph paper vertically represents 10 per cent of a square yard covered by the species mentioned, successive yards being represented by small squares of the paper horizontally. Vertical lines represent single plants. The actual squares have been omitted in the figure for clarity. The slope of the ground is shown in the block at the bottom right hand corner, together with the estimated percentage tree canopy.

layers should also be represented where present, and in other colours.

When the chart is completed the name of the locality and the date should be written at the top and the sheet finally signed by those responsible. Be sure to check the spelling of the Latin names of the plants listed.

Where a quadrat transect has been made, only the positions of the plants in each quadrat are recorded, being taken from the field note-book and entered on a plan drawn out to scale on a large sheet of graph paper. Use symbols to represent the different plants and add a key to these in the margin (see p. 77). Mark clearly the scale of the chart on the paper.

Chapter VII

PROFILE CHARTS AND MAPPING ROOT SYSTEMS

THE *Profile Chart* is intended to give a picture of the composition of the flora from a vertical standpoint, in contrast to the belt transect which gives a plan of the vegetation from a horizontal aspect.

Ideally, therefore, the profile chart should be taken over the same line as that used for the belt transect. In practice this is rarely possible however. In making a belt transect the vegetation is inevitably disturbed, if not to some extent broken down and such an area is quite unsuitable for a profile chart. It is usually found best therefore to make the profile chart along the opposite side of the survey area, thus keeping this work well away from the other groups. If the ecological zones lie across the survey area, with the boundaries of the associations running roughly parallel to the base line, such a chart should give a picture more or less comparable to that of the belt transect, so far as the composition of the vegetation is concerned. This is especially true if the survey is not too large. If a large area is under consideration, or if there is a marked change in the vegetation across the site, it may be better to use one of the parallel grid strings nearer to the belt transect and to carry out additional belt and profile surveys at the other side.

It is rarely possible to obtain an ideal profile chart, for such a chart should include not only a picture of the height and spread of the plants above ground but also of their root systems below ground. It follows, therefore, that theoretically a trench should be dug along the line, sufficiently deep to reach to the bottom of the root systems of the plants involved. Since survey work is generally carried out on other people's land this is naturally

E

impossible, quite apart from the labour involved. More-over in certain types of vegetation, in woodlands for example, the digging of a trench deep enough to reach the bottom of the tree roots would involve labour which is quite impossible to undertake. On the other hand the mapping of root systems is of considerable interest and attempts to do this on a small scale should be made whenever possible. We shall return to this subject later (see p. 71).

The profile chart, as generally made in survey work, is begun by laying out a tape along the grid string, taking care to stretch it tightly, and then proceeding to make drawings of the height and spread of the individual plants, in fact to make a scale drawing of the vegetation as seen in side view. It is rarely possible to do this com-pletely, for there are usually far too many plants and the best that can be done is to choose those plants which occur at selected intervals along the tape. For most types of vegetation the plants occurring at 3 or 6 in. intervals are selected. The height and width are measured with a spring tape, taking care to note the general shape, that is, whether the plant spreads out evenly from the base or whether, like bracken, it begins as a slender shoot and spreads out over other vegetation higher up. The records are made on graph paper in the field note-book and the Latin name of each plant is written beside it.

With the aid of a slight amount of skill and imagination, little conventionalized sketches may be made to resemble the actual plant in general form, but where this is not considered possible symbols may be used instead, with a straight line representing the total height and a line at right angles showing the maximum spread, this line being drawn at the height up the vertical line where the spread is greatest. In some cases it will be sufficient to make a rough outline of the plant to give the general impression of its form, but drawn to scale. The scale selected must vary according to the character of the vegetation, but usually the vertical scale must be very much larger than the horizontal.

The work is slow, if carefully done, and it is best to space out a group of students along the line, giving each person a short length to do, say the plants included in two squares of the mapping grid. Thus if the plants are to be drawn at 3-in. intervals and the grid has been set in 10-ft. squares, it will mean that each worker is responsible for drawing about 80 plants. Care must be taken to see that where symbols are used all the workers are agreed in advance as to the symbols they are to use for the same plant.

On reaching home the results should be set out on the same sized sheets of graph paper as those used in the previous records (see Fig. 21). First the profile of the ground must be drawn from the data given by the levels group. The differences of level should be on the same vertical scale as that to be used for the plants. It will probably be found that on the large scale which is desirable, the whole chart cannot be arranged on a single line across the page, in fact two or even three lines may be drawn one above the other on the same sheet to represent segments of the profile line.

After the line of the ground has been carefully drawn in Indian ink and divided into sections equal in length to the sides of the grid squares, each student should in turn reproduce on the appropriate scale the records made in his field note-book, each plant being placed at the correct distance along the soil line. It must be pointed out that a slight error is almost certain to occur, since usually the plants are measured by a tape stretched along the ground, not by a truly horizontal line. Hence the actual distances measured will be more than those of

FIG. 21.—A profile chart, taken across a Marsh Meadow at Pontfaen, Pembrokeshire. This chart was made during the same survey as that represented in the map in Fig. 13 and was taken along the left-hand margin of the map. The heights of the plants are drawn to scale. The dotted lines across the ditches show the depth of water.

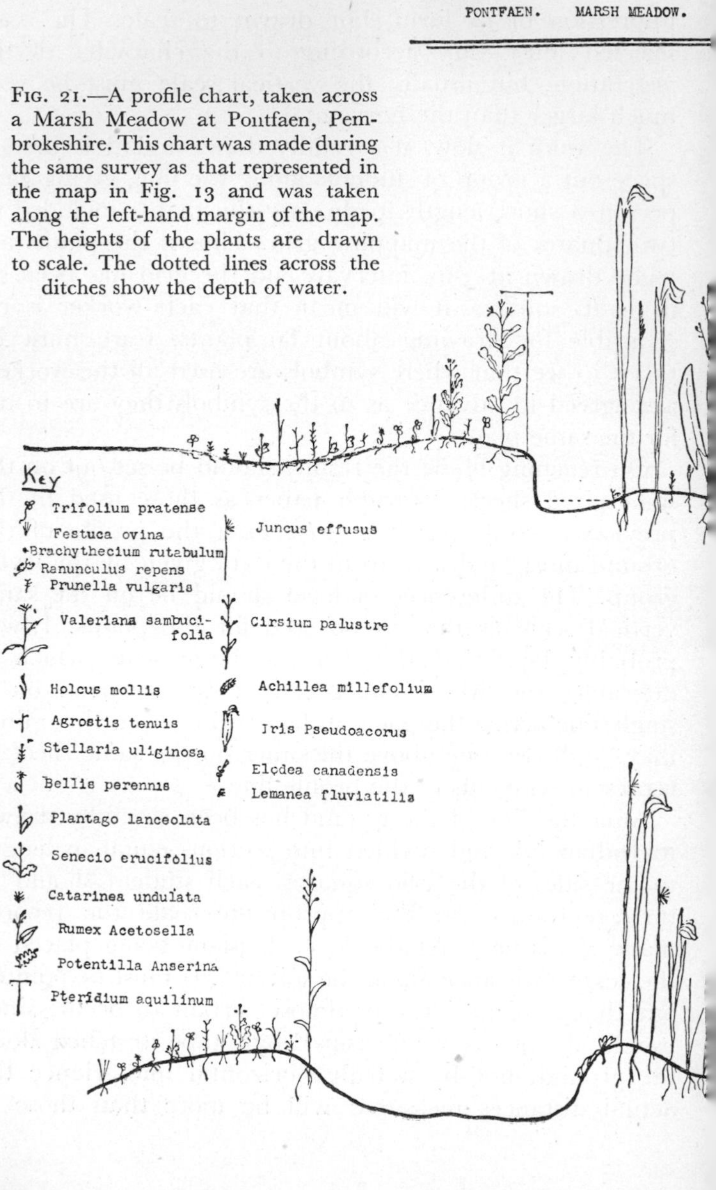
PONTFAEN. MARSH MEADOW.
Key
Trifolium pratense
Festuca ovina
Brachythecium rutabulum
Ranunculus repens
Prunella vulgaris
Valeriana sambuci-folia
Holcus mollis
Agrostis tenuis
Stellaria uliginosa
Bellis perennis
Plantago lanceolata
Senecio erucifolius
Catarinea undulata
Rumex Acetosella
Potentilla Anserina
Pteridium aquilinum
Juncus effusus
Cirsium palustre
Achillea millefolium
Iris Pseudoacorus
Elodea canadensis
Lemanea fluviatilis

June 29th. 1940.

Corylus Avellana

7 FT

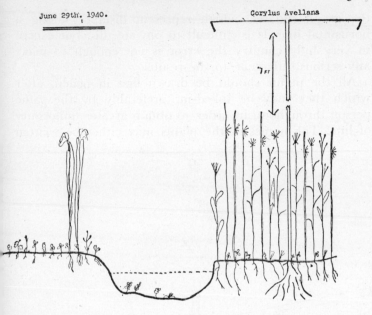

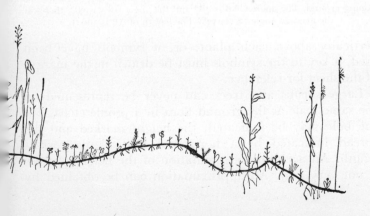

the graph squares, which represent distances along a horizontal line. It is difficult to obviate this, but except in very hilly country the error is not enough to make any serious difference to the results.

All the plants should be drawn first in pencil, after which they may be inked-in, preferably by the same person throughout, in order to obtain greater uniformity of line. The names of the plants may either be written

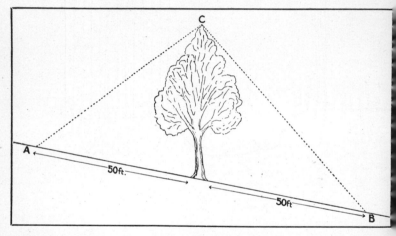

FIG. 22.—Diagram of the method of measuring the height of a tree on sloping ground. The angle CAB is 48° and the angle CBA is 38°, therefore the average angle (α) is 43°. The length of AB is 100 ft.

vertically above each plant, or, if symbols have been used, a key to the symbols must be drawn in the margin of the sheet for reference.

Large shrubs and trees can never be represented on the same scale as the ground flora in a profile chart, but the bole may be indicated, the height marked and the spread indicated by a horizontal line of the correct length. An accurate determination of the height of trees· is not very easy. An approximation can be obtained by measuring a distance of 30 ft. or more from the tree with a tape and sighting the top with a clinometer compass from this point (see Fig. 22). The angle between the ground and the line of sight to the top of the tree is recorded. The height is calculated from the formula:

$$X = L \times \tan \alpha,$$

where $X =$ the height of the tree, $L =$ the distance of the compass observation from the bole, and $\tan \alpha$ the tangent of the observed angle. This calculation assumes that the ground is approximately level. On sloping ground the angle between the tree and the ground is not a right angle and the resulting height is either greater or less than the actual. On a hilly site, therefore, if a level observation is impossible, the same distance should be measured out above and below the tree and the angle taken at equal distances on either side. The average of these two angles gives the true angle on a horizontal site.

When all the data have been put into the chart, the locality and date are written at the top of the page and the sheet is signed by the members of the group in the bottom right hand corner.

Maps of root systems are sometimes possible, but only in fortunate situations unless, as has been said above, a trench can be opened. The side of a quarry may provide a good opportunity, or good root systems may be found in sand dunes, where the loose sand can be carefully removed from the plants to expose the roots. Although such plans will not usually form part of a particular survey, they may be made to provide additional information.

Where the root system is comparatively small, as for example in *Calluna vulgaris*, it may be mapped by freehand as shown in Fig. 23. If, however, a more accurate diagram is desired, take four yard-sticks and tie them together to make a 2-ft. square. Stretch thin string across

Fig. 23.—Mapping the root system of a plant of *Calluna vulgaris*, on Studland Heath, Dorset. The sand has been carefully cleared away from the side of the sand dune to expose the whole of the root system. The string grid used to give the scale has been removed to show the actual roots.

to divide the square into either a 1-in. or 3-in. mesh according to the amount of detail desired and arrange the screen vertically against the root system. Choose a scale in the field note-book suitable to the size of the root system and draw the system in as accurately as possible, making use of the screen to keep the scale right. It will be found easiest if each square of the mesh is drawn in turn starting from the ground level. Arrange the screen so that the top is horizontal, using a spirit level or a plumb line to ensure this, and then first draw in the ground level, showing any slope which occurs. If the system is too large to be covered at one time by the screen, mark the corners with quadrat spikes before moving the screen to the next position.

Wherever possible adjacent plants, such as small grasses, should also be considered. Observe and record any vertical zoning of different root systems. For example, in the case cited above, it will be seen that the root system of *Calluna* usually spreads out well below the ground level, leaving an upper zone about 6 ins. deep which is occupied by the roots of small grasses. If bracken is also present its rhizomes may be found growing horizontally below the root system of the *Calluna*. All these points should be brought out in the plan.

At home the plan may be re-drawn on a larger scale and the root systems of the plants drawn in Indian ink.

If the side of a quarry which is being studied, includes the root system of a tree, the small screen suggested above may be too small. In this case tapes must be used, one being stretched along the surface of the ground, the other hung downwards. Measurements from these tapes are then made with yard-sticks or spring tapes to help in making a scale drawing. It is very difficult to get a root system of this size accurately represented and photography is often the best way to bring out the required detail. If a photograph is taken, remember to include in the picture a clear scale, say a yard-stick, to indicate the actual size on the final print.

Chapter VIII

QUANTITATIVE AND SOCIOLOGICAL ANALYSIS OF VEGETATION

GREAT importance should be attached to the analysis of the vegetation, although in the past this aspect of survey work has been much overlooked. Such an analysis may be either quantitative or sociological. In the former a quantitative estimation of the species composing each association is made, either from selected sites or by random sampling. In the sociological analysis the relation of the plants to one another and to the habitat is investigated.

The oldest, though possibly not the best, method of quantitative analysis is the *Quadrat*. A typical area of the survey is selected, in which a careful analysis of the plants is to be made. The size of the quadrat must vary according to the nature of the association. In a vegetation composed of small plants, such as a meadow or a marsh, a square yard is usually chosen. In a wood a considerably larger area must be employed, since otherwise the trees and large bushes will be excluded from the survey. Here a 10-ft. grid square is often taken, though occasionally an even larger area may be necessary. Detailed quadrats on this scale take much time, usually more than the results justify, and, where students are concerned, such analyses are best confined to the ground flora in open situations.

Suppose a grassland is under survey. Select a site which on inspection appears to be typical of the area and put in four spikes to mark off a square yard, taking care to get the corners true right angles. For this two spring tapes may be used in the way described in mapping (see p. 37). Join the four spikes with a piece of string.

Locate the quadrat by measuring from the grid strings so that its position can be recorded on the map.

Now divide the square yard with strings into four equal squares. Four students can now set to work on the area, each working from a different side. Working with graph paper in the field note-book, the position and spread of each plant is carefully marked. Symbols may be used, but if so each member of the group must use the same symbols.

On reaching home the four quarters are mapped together on a single sheet of paper (see Fig. 24), and the whole quadrat drawn out on a large scale. A key giving the names of all the plants recorded is placed at the side of the chart. Such work needs the careful use of a Flora and possibly the assistance of someone who has a good knowledge of native plants, for it must be remembered that *all* plants have to be identified, not only those in flower. Keys to the identification of British grasses by their vegetative characters have been compiled and will be found of particular assistance in these studies (see Appendix I, p. 181).

Several quadrats should be studied in any one survey area, in fact two quadrats in each community are desirable. This may mean that possibly a dozen quadrats should be studied on a single survey site.

Permanent quadrats are of considerable interest. After the quadrat has been mapped, permanent pegs are driven into the ground in the holes previously occupied by the spikes. If permanent corner pegs have been set up to mark the whole survey area, it should be possible by careful measurement to locate quadrats again. If visited after a period of years the comparison of the two maps may bring out interesting information. Changes in the composition of the quadrat may be observed, certain plants have increased, encroaching on others which have

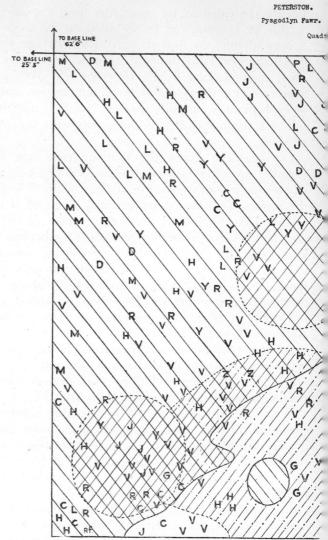

Fig. 24.—A typical Quadrat, made in a bog at Pysgodlyn Fawr, Peters
tion of the Quadrat in relation to the survey Base Line is marked in the
by permanent pegs so that the actual Quadrat could be located for

pril 23rd, 1938.

B.

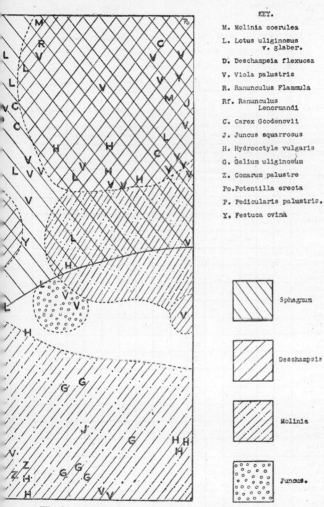

KEY.

M. Molinia coerulea

L. Lotus uliginosus
        v. glaber.

D. Deschampsia flexuosa

V. Viola palustris

R. Ranunculus Flammula

Rf. Ranunculus
        Lenormandi

C. Carex Goodenovii

J. Juncus squarrosus

H. Hydrocotyle vulgaris

G. Galium uliginosum

Z. Comarum palustre

Po. Potentilla erecta

P. Pedicularis palustris.

Y. Festuca ovina

Sphagnum

Deschampsia

Molinia

Juncus.

organ. The key to the species is given on the right. Note that the posi-
and corner. In this survey the position of the Base Line was marked
ive study at a later date.

BOSSINGTON.   SOMERSET.   JUNE – JULY.   1939.

VALENCE ANALYSIS.

Student: P. Burgoyne   Date: June 26th.

Sheet No.1.

Area surveyed: South Bank,River Barle. Simonsbath.Devon.

Type of Community: Molinia-Sphagnum Bog.

Size of Trial Areas: 6 inches square.

Species	Numbers of Trials														
	1.	2.	3.	4.	5.	6.	7.	8.	9.	10.	11.	12.	13.	14.	15.
Molinia coerulea	–	–	–	–	–	–	–	–	–	–	–	–	–	–	–
Sphagnum sp.	–	–	–		–	–		–		–		–	–	–	–
Potentilla erecta	–			–	–			–	–				–		–
Narthecium ossifragum	–							–			–				
Festuca ovina		–					–	–	–					–	
Ranunculus Flammula		–													
Erica Tetralix			–	–	–						–	–	–		
Juncus effusus			–								–				
Luzula pallescens			–												
Eriophorum vaginatum				–					–						
Luzula pilosa					–										
Scabiosa succisa						–			–						
Viola palustre						–								–	
Wahlenbergia hederacea						–								–	
Drosera rotundifolia							–								
Polytrichum commune									–						
Carex flava										–					
Pinguicula lusitanica										–					
Vaccinium Myrtillus												–	–	–	–
Hypericum elodes												–			
Pedicularis sylvatica													–		
Carex binervis															–

FIG. 25.   *For caption see opposite*

decreased accordingly. Thus, features of the competition between species and trends of change in the development of the association, which may not be noticed except by this method, are well brought out.

When a survey area is easily accessible for visits at short intervals another form of quadrat may be tried. This is the *Denuded Quadrat*. A square is marked off and analysed as before. Then every plant, large and small, of whatever character, is carefully removed, roots and all, leaving the soil bare. The object of this proceeding is to provide a free, if not quite virgin habitat, in which the processes of colonization and succession can be observed at work gradually re-building the association which surrounds the area, or, it may be, deviating from it towards the development of a somewhat different community. As an object lesson in dynamic ecology it is unique and it is a great aid to students in picturing the long process of succession by which stable types of vegetation such as we find in nature, have arisen.

Valence Analysis is an analysis of the percentage occurrence of species composing random samples of the vegetation. As with the quadrat, the size of the area subjected to a valence analysis may vary from a 6-in. square to one of several feet, according to the nature of the vegetation, but the most interesting analyses are usually obtained on land with a dense ground flora, such as a moor, a meadow or a marsh. In such localities with many small plants, a 6-in. or a 1-ft. square is suitable. Square frames made of half-inch by one-eighth inch iron strip can be used, as already described (see p. 24), or a

---

The working sheet of a Valence Analysis, made at Simonsbath, Devon, in a Sphagnum bog near the River Barle. The names of species have been typed in for clarity. A horizontal line indicates the presence of the species in each trial area.

peg with a length of string attached may be used as follows. Cut short lengths of thin rope or cord, turn one end over to make a loop and bind it down. Cut the rope into such a length that it forms the radius of a circle of definite area, including the loop at the end in the length. For example, a rope 6·5 ins. long is approximately the radius of a circle 1 square foot in area. Peg the rope down with a stick through the loop and by rotating the free end the required area is obtained.

The method of sampling is to toss the square down on the ground, preferably over one's shoulder, to avoid the element of choice, and work the area where it falls. In the case of the peg and string, the peg is thrown down at random and pushed in where it falls.

Special ruled sheets of paper are generally used for this work (see Fig. 25). In the left-hand vertical column are written the names of the species found, while in the top horizontal row are given the numbers of the trials made. At least twenty-five trials are usually made in each association. Two sheets will suffice for each association, unless the number of species found is very great. Some workers prefer the selection of "typical" areas in the association for valence analysis instead of the random sampling here suggested, on the ground that a smaller number of samples are required. This, however, presupposes considerable experience to guide the selection, and for beginners, at any rate, the random method is safer, provided as many samples as possible are taken. Each species occurring within the area of the square is marked in the appropriate space by a line.

Care must be taken to see that the trials are all conducted within the same plant association, for if sampling is extended over several different associations the results will be quite anomalous and misleading.

The results of a valence analysis are worked up in the

following way. The total number of occurrences of each species is counted up and expressed as a percentage of a hundred trial areas; i.e. the number of occurrences in the twenty-five areas is multiplied by four. The species are then grouped according to the values found for the percentage number of occurrences. Five groups are taken as follows:

Group I	..	..	.. 0–20 per cent
Group II	..	..	.. 21–40 per cent
Group III	..	..	.. 41–60 per cent
Group IV	..	..	.. 61–80 per cent
Group V	..	..	.. 81–100 per cent

When this has been done the fraction of species occurring in each group is calculated in terms of the total number of species recorded. For example, if thirty species were recorded in the trials, of which three were found in over 80 per cent of the trials, this would mean that one-tenth of the species occurred in the fifth group. This is then expressed as a percentage of the total number of species, i.e. 10 per cent.

A histogram is now made (see Fig. 26) by dividing the horizontal scale into intervals representing the five groups and dividing the vertical scale to represent the percentage of species occurring in each group.

The histogram so formed gives a clear idea of the composition of the vegetation and, in a typical association, generally assumes the shape of a capital **J**, where the largest number of species occurs in the smallest percentage group. In other words, only a small number of species occur in all, or nearly all, the trials, while a large number of species occur only sporadically. Those species occurring in 80–100 per cent of the trials may be regarded as the dominants and co-dominants, while those found in the first group may be looked upon as accidentals.

F

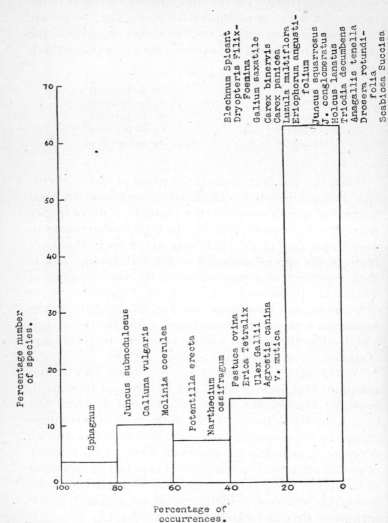

VALENCE OF MOORLAND VEGETATION ON CARN INGLI
Newport, Pembrokeshire.
Sphagnum Area.   June 30th. 1940.

FIG. 26.   *For caption see opposite*

The *Abundance* of a particular species may also be determined by counting the number of plants of that species in each trial area. The same areas can be used and the estimation can be done at the same time as the valence analysis. In the case of creeping or tillering plants, e.g. grasses, a count of the number of shoots coming through the ground may be used, instead of a count of actual individuals, which may be impossible to ascertain.

The results of these two estimates are not necessarily similar, for it does not by any means follow that because there are a large number of individual plants in a particular area that it has the highest valence, since the latter depends on uniformity of distribution as well as on abundance.

The results of this series of observations cannot be easily expressed graphically, and it is best to set out simply a list of species arranged in order of abundance. This may be done in the following way. For each sample area calculate the percentages of the individuals of each species present: e.g. among the individual plants present in a single sample there were—*Bromus erectus* 80 per cent, *Poa pratensis* 6·6 per cent, *Trifolium procumbens* 5·9 per cent, etc.

The abundance of a species in the association is the average of these percentages over the whole series of samples. Alternatively the species may be arranged into five groups in decreasing order of abundance, in the same way that the valence analysis groups them in order of frequency.

Valence Histogram, compiled from data recorded as illustrated in Fig. 25, made in a Sphagnum bog on Carn Ingli, Newport, Pembrokeshire. The vertical scale shows the percentage number of species and the horizontal scale the percentage number of occurrences. Names have been typed in for clarity.

The *Covering Capacity* or *Cover Index* of a species has already been referred to in connection with the belt transect (see p. 59). Similar observations may be made on the same trial areas as those used for valence and abundance studies. In this case the percentage of the surface of each trial area covered by each species is determined by inspection. The results are calculated by estimating the percentage area occupied by each species in one hundred trials and arranging the results in a list of decreasing percentages.

A comparison of these three observations is now possible by listing, in parallel columns, the percentage frequency or valence, the percentage abundance, and the percentage covering capacity of the chief species in the association or their group numbers under each heading, if the group classification is adopted. Taking these three results together, clear indications of the average composition of the association may be obtained. These results, together with the valence histogram, should be set out on a large sheet of paper. The name of the locality and the date is written at the top, the sheet being finally signed by the members of the group which did the work. Four students are generally employed as a group for this work, though on a large survey with a number of different associations, a second independent group can be conveniently employed.

The example on page 85 may be quoted to show the kind of results which may be expected.

By the term *Sociability* is implied the condition in which the individuals of any species occur in relation to one another, that is to say, whether they occur as scattered individuals or in a number together. This is an intrinsic quality of the species and is therefore independent of the quantitative factors of frequency, cover-index and abundance, which characterize the species in relation to other species making up the community and are, therefore, as we have suggested, comparable characters, which are best discovered by the analysis of random samples.

Sociability is a character quite independent of the

RODBOROUGH COMMON, GLOUCESTERSHIRE, APRIL 1942

Comparison of Valence, Abundance and Cover Index

Species	Valence	Abundance	Cover Index
Bromus erectus	100	76·95	52·56
Poa pratensis	92	8·94	5·44
Vicia hirsuta	68	2·20	2·16
Fragaria vesca	60	2·14	4·28
Taraxacum officinale	56	1·44	2·36
Carex flacca..	44	1·15	1·36
Anthyllis vulneraria	40	2·49	4·12
Digitalis purpurea	40	0·79	2·16
Epilobium angustifolium	32	0·52	1·56
Dactylis glomerata	16	1·50	1·60
Polygala vulgaris	16	0·73	0·20
Rubus saxatilis	16	0·30	0·88
Lotus corniculatus	16	1·65	1·12
Veronica officinalis	12	0·24	0·08
Vicia sativa	12	0·18	1·20
Hypochaeris radicata	12	0·18	0·20
Trifolium procumbens	8	0·36	0·40
Helianthemum Chamaccistus ..	4	0·06	0·02
Hieracium Pilosella..	4	0·14	2·36

Average of bare ground, 19 per cent. Average number of plants in each 6-in. square, 77·64.

above. It can be estimated arbitrarily by inspection of the area as a whole and the comparison of the chief species. This inspection may follow the completion of the previous analyses, when it should be evident to the student which are the principal species of the community, even before the detailed results have been worked out.

Make a list of, say, a score of such species. Compare them with each other from the point of view of Sociability, placing them in five groups, of which the first will contain the species that are more or less uniformly dispersed and the fifth will contain those that show the most marked degree of aggregation into clusters, that is, which depart most from a true random distribution. Such an assessment should not be beyond the student's capacity, and although it is only semi-quantitative it provides a type of information which may be, for example, in cultivated grassland, of considerable practical importance. It is susceptible of accurate statistical treatment, but this is beyond the scope of these elementary studies.

The *Stratification* of the vegetation is another sociological factor which can be noted. Reference to this has already been made in connection with mapping and transects. It was pointed out that the spread of trees should be dotted in on the map since this may play an important part in determining the composition of the ground flora. In connection with belt transects we saw that the covering capacity of the vegetation must be treated separately in so far as the various strata were concerned, and the same treatment was advised in profile charts. In a small area no further consideration is possible, though it is worth noting that some ecologists maintain that since many types of vegetation possess several obvious strata each should be regarded as a distinct community and treated separately so far as the analysis of the vegetation is concerned.[1]

[1] Two other sociological characters which have been proposed are excluded here because they require a comparative knowledge of plant communities which is beyond the average student. These are Association Constancy, or the regularity with which a species appears in the same association in different areas; and Association Exclusiveness, or the degree to which a species is confined in its occurrence to one particular association. See Braun Blanquet: *Plant Sociology*.

In 1907 Raunkaier published his description of a series of *Life Forms*, which has a decided bearing on the analysis of vegetation. As a result of his researches he was able to recognize the following types, based upon the mode of continuance of the plants from year to year.

1. *Phanerophytes.* Plants whose buds and apical shoots survive unfavourable periods of the year projecting in the air on stems which live for several or many years. Such plants have stems which are negatively geotropic, so that each year the buds of the subsequent year are carried further from the ground.

2. *Chamaephytes.* Plants whose buds and shoot apices destined to survive unfavourable periods of the year lie either on the surface of the ground or quite near to it. The flowering shoots are negatively geotropic while frequently the persistent shoots lie along the surface of the ground or do not project more than 30 cms. from it.

3. *Hemicryptophytes.* Plants in which the shoots die back to ground level at the beginning of an unfavourable period, so that only the lowest portions remain alive, being protected by the soil and withered leaves. The buds on these portions survive the winter in the surface layers of the soil and in spring grow up to produce shoots bearing leaves and flowers.

4. *Cryptophytes.* Plants whose buds or shoot apices destined to survive unfavourable periods are buried in the soil or else lie at the bottom of water, the depth varying in different species.

5. *Therophytes.* Plants which survive the unfavourable periods of the year simply as seeds. Such plants pass through their life-cycle in a few weeks or months and correspond roughly to the "annuals."

In any sample of vegetation it is obviously possible to divide up the plants according to this classification. In this way is obtained what Raunkaier called the *Biological Plant Spectrum*. A spectrum may be made of the composition of the flora of a country or of a continent, and such spectra show remarkable differences according to the major climatic conditions. Thus continental countries in the higher latitudes show a greater proportion of Hemi-cryptophytes and Cryptophytes than of Phanerophytes and Chamaephytes, in comparison with a tropical country where the reverse will be true.

A similar spectrum can also be made of a plant community or plant association, and interesting results can be obtained which give some precision and a quantitative character to the idea which the older ecologists called the "physiognomy" of a plant community. Such spectra are most conveniently based on a valence analysis of the vegetation.

The following examples will show the kind of contrast which may be expected from analyses of respective associations.

Raunkaier's Life Forms	Chalk down, Swanage, 1938	Mixed woodland, Bossington, 1939	Bog-land, Exmoor, 1939	Sandy heath, Studland, 1938
	per cent	per cent	per cent	per cent
Phanerophytes ..	3	28	8	55
Chamaephytes ..	32	6	21	11
Hemicryptophytes ..	55	41	58	11
Cryptophytes ..	0	19	13	23
Therophytes ..	10	6	0	0

An interesting practical application of ecological survey methods is the procedure known in Sweden as "Taxing the Forests." It is carried out, as a rule, every ten years on each forest property. The object is to ascertain the total quantity of timber standing and its annual rate of increase.

First of all, lines 100 metres apart are ruled on a map of the forest. The party then goes to the starting point of the first line, where one man ties a rope round his waist, takes a compass bearing, previously ascertained from the map, and starts off through the forest, following this line as closely as possible, while the rope trails after him to mark his route. He is followed by two others with a steel tape, who measure off along the rope the longitudinal distance traversed.

Another pair, carrying 5-metre poles as guides, measure with calipers the breast-high diameter of all trees within 5 metres of the rope on each side. They call out the species of tree and the diameter in centimetres to a sixth man, who notes them down. By following each line in this way, right through the forest, 10 per cent of the trees are individually measured.

Every 500 metres the party halts and a quadrat of .10 by 20 metres is marked out. In this every tree is measured and a small core is bored out, in which the age of the tree is counted from the annual rings, the number of rings in the last centimetre of wood are counted and the thickness of the bark is measured. The height of each tree is also measured with a sighting clinometer. Thus all the factors are obtained for calculating the total volume of timber on the area and its percentage annual increase.

The party is accompanied by a surveyor, who marks on a sketch map the topography of the ground and the "class" of forest in different areas. The "class" is determined by the prevailing species and the average age of the trees.

He also estimates arbitrarily the "bonitet" of the soil, that is its presumable productive power in cubic metres of timber per annum, and the "status" of the forest, which means its character judged by a theoretically perfect stand of timber, which is classed as 1·0. If the forest is too thin the status is below unity, e.g. 0·5, etc. If the ground is overstocked the figure is above unity.

Chapter IX

PEDOLOGY: EXAMINATION OF THE SOIL

THE character of the soil in which plants grow plays a very important part in determining the character of the vegetation. It follows therefore that a study of the soil forms an essential part of plant ecology. Too often this is forgotten and the nature of the soil receives scant attention.

While it is not possible to carry out a complete analysis of a soil in the field, certain important features can be investigated, indeed can only be investigated there. Extended tests can be undertaken independently on samples in the laboratory. Certain of these are essential for an understanding of the soil and are dealt with here; other and more detailed studies are outside the scope of this book.[1]

Some of the following observations should be made preferably in each square of the grid, or at least in each plant community within the survey area.

Soil Profile

With the aid of a spade dig a hole sufficiently deep to pass well down into the sub-soil. Take care that at least one side is truly vertical and make a plan to scale in the field notebook, of the successive layers. Good soil profiles are difficult to obtain in very dry or loose soil, but enough can usually be seen to learn how the quality of the profile varies in different areas.

The classification of the soil layers in a profile has been very critically studied by experts who have come to recognize a number of characteristic layers. The following

[1] Students requiring more detailed instruction in the analysis of a soil are referred to our *Text-book of Practical Botany* (in the Press).

table shows the layers which are found in the fully developed profile of a highly leached soil or Podsol. In immature soils or soils in dry situations, some of the zones may be absent or poorly developed, but the following may be mentioned as those typically recognizable.

Horizon A.

Zone A, oo. Sub-layer L. Undecayed surface litter.

Zone A, o. Sub-layer F. Material still recognizable as to its origin.

(These sub-layers are most easily distinguishable in woods.)

Zone A, 1. Upper layer of mineral soil with humus incorporated.

Zone A, 2. Layer in which leaching is usually evidenced by the pale colour. In extreme cases the layer may be grey or even white.

Horizon B.

Zone B, 1. Layer in which humus leached from above is accumulated to form a dark "pan."

Zone B, 2. Layer in which the iron sesquioxide leached from above is accumulated to form a dark brown layer.

(This horizon is absent from soft calcareous rocks.)

Horizon G.

Gley Zone. Greyish zone with greenish or yellowish mottling.

(This layer is chiefly found on boggy ground and is supposed to indicate periodical flooding of the soil.)

Horizon C.

The parent rock material, practically unweathered.

Even where the zonation is poor, the main Horizons, A, B and C can usually be distinguished.

Some indication of the nature of the sub-soil may also be obtained by a study of the geological map of the district, but it must be remembered that such maps, though in the main accurate, do not take into consideration minor variations in the nature of the sub-soil, and the actual material on a particular site may not be correctly represented on such maps. In the case of transported soils, such as alluvium and the widely distributed glacial or boulder clay, horizon C is not, of course, the parent material of the upper layers and there is no con-

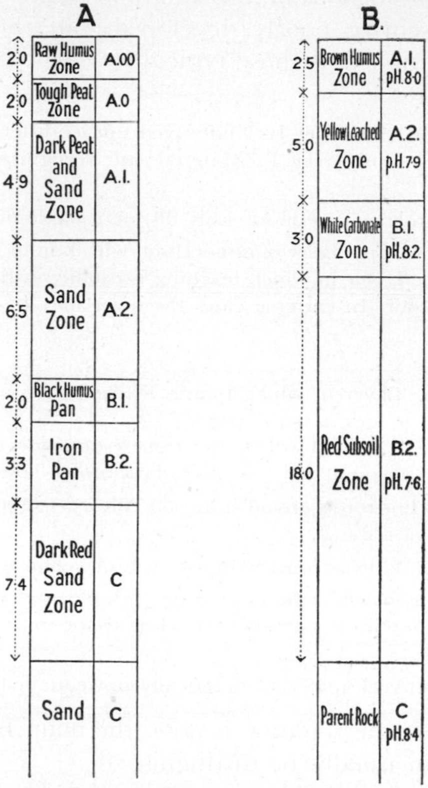

FIG. 27.—Two Soil Profiles. A, represents a strongly podsolised soil from Cavenham Heath, Norfolk, made by Dr. E. P. Farrow. The leached "A" zones, and the "B" zones of accumulation are well contrasted. B represents a soil of Rendzina type from under grassland at Rodborough Common, Cranham, Gloucestershire, made by Miss E. Moore. Depths measured in inches.

tinuity between it and them. Nevertheless some of the effects of leaching into differentiated layers may be seen even in such soils.

Fig. 27 illustrates sample profiles of a typical Podsol and a Rendzina. The latter name is applied to humus soils containing an excess of free calcium carbonate. In this country such soils may occur on chalk and limestone.

Sampling

To make a good sample of the soil to be studied it is necessary first to clear away the ground vegetation together with an inch or two of the semi-decayed humus. This roughly corresponds to the sub-zones A, oo and A, o of the profile. The soil-borer is now thrust into the soil, using the wooden handle to help to turn it. The depth to which the borer is thrust down will depend to some extent on the nature of the vegetation, as well as on the soil itself. If it is desired to pay attention mainly to the effect of the soil on the ground vegetation, which is as a rule shallow rooted, a sample from between 2 and 12 inches deep should be taken.

If the soil is very hard the auger borer may be substituted. A plate placed around the top of the borer, as previously described, will catch the crumbs as they fall from the flanges of the tool.

If the general character of the soil of an area is to be studied, the samples taken from at least six different spots should be mixed, otherwise each sample should be put in a separate soil-tin. The position of the sample on the grid should be recorded in the field notebook together with the number of the soil-tin in which it has been placed. This method should be employed whether the sample is to be roughly analysed on the spot or taken to the laboratory for a more detailed investigation. The depth at which the sample has been taken should be accurately recorded by measuring the hole with a yardstick after the borer has been removed.

Soil Reaction

The Soil Reaction is measured in the field by means of indicators. Soils are naturally so well buffered that considerable amounts of water may be added without altering the pH, so that it is not necessary to weigh out any fixed amount of soil for the test.

A small portion is placed in a test tube and shaken up with distilled water. Allow it to settle and decant off a little of the more or less clear fluid. Test this with a drop of either Universal Indicator, or with the B.D.H. "4–11" or "Soil" Indicators. The colour produced will give a rough indication of the pH of the sample.

If the soil suspension refuses to settle quickly a clear liquid may be obtained by adding to the soil in the test-tube an equal amount of precipitated Barium sulphate and shaking up. The Barium sulphate will often carry down with it, in settling, the fine colloidal matter in suspension.

For more accurate measurements the B.D.H. "Capillator"[1] is recommended. A soil suspension is made as above and a drop is drawn up to a fixed mark in a capillary tube and discharged into a small watch glass. Then a drop of the most suitable indicator for the pH of the sample is drawn up in the same capillary, mixed with the drop of soil suspension, and the combined drops are drawn back again into the capillary. The colour of the indicator is then compared with a series of standard capillaries mounted on a card. By this means the pH can be estimated to 0·1 units.

The following are the most suitable indicators for the purpose and the range of pH over which they are most efficient.[2]

[1] See Chapter II, p. 30.
[2] Messrs. The British Drug Houses, Ltd., supply a compact case containing a set of the capillator cards and the indicators listed, which is very useful both for soil and water analyses.

Indicator	pH	Colour change
Bromophenol blue ..	2·8–4·6	Yellow to violet
Bromocresol green ..	3·6–5·2	Yellow to violet
Methyl red	4·4–6·2	Red to yellow
Bromocresol purple ..	5·2–6·8	Yellow to violet
Bromothymol blue ..	6·0–7·2	Yellow to blue
Phenol red	6·8–8·4	Yellow to red
Cresol red	7·2–8·8	Yellow to violet-red
Thymol blue	8·0–9·6	Yellow to blue
B.D.H. "Soil" indicator	4·0–8·0	Red through green to blue
B.D.H. "4–11" indicator	4·0–11·0	Red through green to violet
B.D.H. "Universal" indicator	3·0–11·0	Red through green to violet

The following are examples of actual pH measurements made in the course of survey work:

Type of soil	Depth of layer	pH
	inches	
Clay; Lisvane, 1938..	4	5·4
Beechwood: Cranham, 1942	4	7·6
Limestone meadow; Dinas Powis, 1942 ..	4	5·8
Peat bog; Margam, 1938	4	4·2

Carbonate Content and Lime Deficiency

Take a few grams of the soil and test with a few drops of 20 per cent Hydrochloric acid. If there is any fizzing it shows that carbon dioxide is being liberated from free carbonates present in the sample. Purely arbitrary grades of fizzing from 1 to 10 have been distinguished to indicate the relative amounts of free carbonate present.

If there is no reaction with acid, it probably indicates a

deficiency of lime, and the soil sample should be tested in the following way using Comber's Thiocyanate Test.

The reagent is made by saturating absolute alcohol with Analar Ammonium thiocyanate. Dry soil must be used and added to the solution in a test tube. The development of a red colour indicates a lime deficiency and the depth of colour is a roughly quantitative index of the amount of this deficiency. If the reaction is weak the colour may take some minutes to appear. This is really a test for ferric ions and is based on the fact that only when there is a deficiency of bases will there be any excess of ferric ions to react with the thiocyanate. Thus the deeper the red colour produced the greater the deficiency of bases present.

A modification of this test, which may be applied as an alternative, is to use a 5 per cent aqueous solution of Potassium salicylate, which gives a violet colour with ferric ions. The chief advantage of this test is that it can be applied to a wet soil. The colour is less intense than with Ammonium thiocyanate.

It may sometimes happen that soils in which reducing conditions prevail, particularly boggy soils, may give no reaction with the thiocyanate even though there is a considerable base deficiency. This is because the iron present is in the ferrous state. If there is any suspicion that this may be the case, add a few drops of Hydrogen peroxide after the thiocyanate reagent. The red colour due to the ferric ions will then appear.

The authors have found that another test for lime deficiency which is even more sensitive than the above is provided by Thioglycollic acid or Sodium thioglycollate.

Mix a small quantity of the soil with water in a test tube, add a drop or two of strong Ammonia and shake. Allow to settle, then add one drop of Thioglycollic acid. A strong purple colour develops in the presence of free

Iron ions, either ferric or ferrous, thus implying, as pointed out above, a deficiency of base in the soil.

Natural Water Content

The absolute water content is a very fluctuating quantity depending on the recent rainfall. An average estimate can be obtained when the sample is taken about 24 hours after rainfall. In the survey of a limited area, however, it is often more important to know the relative moisture of different parts of the area. This can be done by comparing samples taken simultaneously at any time, except during heavy rain, when the soil may be saturated.

Take a sample of about 10 gms. of soil and weigh it in a weighed basin. Dry in an oven at or near 100° C. Cool in a dessicator and reweigh. Replace in the oven again for an hour, and continue to do so until no further loss in weight occurs. Care must be taken not to heat the soil too greatly or weight losses will occur due to changes in its composition.

Calculate the loss in weight due to water as a percentage of the weight of oven-dried soil.

Saturation Capacity

This may be defined as the percentage of water held against gravity by a soil in which the pores are completely filled with water.

The measurement is made with a "soil box" (see Fig. 28). This can be made from a piece of zinc tubing 2 ins. in diameter and about $1\frac{1}{2}$ ins. long. Inside, near one end there fits a stout brass wire ring, to support a disc of perforated zinc which has been carefully cut to fit inside the tube. This perforated bottom is held in position by a second brass ring. A few holes are drilled in the lower end to admit water.

Fit a piece of filter paper over the perforated bottom

G

and weigh the whole. Now fill the box with air-dried soil which has been thoroughly broken up and sieved through a 1 mm. sieve. Carefully pack the soil into the box, tapping it on the bench, while filling, to let it settle thoroughly. Finally with a ruler strike off the soil flush

Fig. 28.—Two types of "Soil Boxes." On the left is a shallow form suitable for samples of approximately 200 gms., while that on the right is more suitable for small samples. The latter has been turned upside down to show the holes through which the water enters. The perforated zinc lies across the centre of the cylinder.

with the top of the box. Weigh again, thus obtaining the weight of the soil contained in the box.

Place the box in a flat trough with enough water to soak into the soil but not enough to come over the top of the box. Leave until the surface of the soil appears thoroughly wetted.

Dry the surface of the box rapidly and allow the few

drops of water which come from the bottom to drain away. Weigh again. This gives the weight of the water absorbed into the soil.

Now place in an oven at 110° C. and dry off again, cooling and re-weighing, until a constant weight is obtained. Finally, determine the weight of water held by a piece of filter paper of similar size and texture to that used in the bottom of the soil box.

If $a =$ weight of box and filter paper,

$b =$ weight of box, filter paper and air-dried soil,

$c =$ weight of box, wet filter paper and wet soil,

$d =$ weight of box, dry filter paper and oven dry soil,

$f =$ weight of water held by the filter paper,

then the moisture content of the saturated soil, i.e. the weight of water in grams taken up by 100 gms. of oven-dried soil equals

$$\frac{[c - (a + f)] - \{b - [a + (b - d)]\}}{b - [a + (b - d)]} \times 100$$

Which may be simplified in practice to:

$$\frac{c - (d + f)}{d - a} \times 100$$

Oven-dried soil is often difficult to wet. It is therefore important to start analyses with air-dried soil.

A few examples of actual saturation capacities determined are given below:

Kind of soil	Saturation capacity of 100 gms. of dry soil
	per cent
Clay soil; Lisvane, 1938	58·4
Beechwood; Cranham, 1942 ..	69·7
Limestone meadow; Dinas Powis, 1942	70·0
Peat bog; Margam, 1938	95·2

Sticky Point

A determination of the "Sticky Point" of a soil is a very useful, and at the same time simple, method of characterizing its physical nature. At the sticky point the soil materials are reduced to their condition of closest packing, the colloidal material is fully saturated and all air has been expelled. The sticky point is, therefore, a good index of soil texture for comparative purposes.

Spread out, on a glass plate, about 10 gms. of dry soil which has been shaken through a 1 mm. sieve. Gently spray it with water from a wash bottle. Continue this,

Type of soil sample	Weight of water absorbed by 100 gms. of oven dry soil
	per cent
Clay; Lisvane, 1938	33·7
Beechwood; Cranham, 1942	42·2
Meadow on limestone; Dinas Powis, 1942	44·0
Peatbog; Margam, 1938	71·5

mixing with a spatula, until the soil is wetted and becomes sticky. Then work the mass up with the spatula and finally scrape the soil off the glass and knead it in the hands until a point is reached at which the mass no longer spreads out but begins to clean up the fingers and form a smooth ball. At this point it can be cleanly cut with a knife. A sample of the ball is now weighed, and then dried in an oven at 110° C. Continue heating, cooling and weighing until a constant weight is obtained.

The difference in the two weights gives the amount of water necessary to saturate fully the colloidal soil particles and fill all the interspaces with water. The above figures of the same soil samples as in the previous table will give an indication of the variations likely to be obtained. They are expressed as a percentage of 100 gms. of oven-dried soil.

It has been found, as a result of experiments on soils in various parts of the world, that there is a relationship between the maximum water holding capacity of a soil, that is its saturation capacity under natural conditions, and the sticky point, and a constant has been devised which relates the two estimations. Though experts differ somewhat in the value they give to this constant, we may take the figure of 23.5 as most suitable for our purpose.

Thus in the case of the clay soil from Lisvane, Glamorgan, the analysis of which is given above, it will be seen that the sticky point determined was 33·7 per cent. Adding the above constant, the water-holding capacity of the soil would be 57·2 per cent, while the result obtained experimentally came to 58·4 per cent, which is within the error to be expected in soil estimations.

Oxidizable Organic Content

It is not easy to obtain an accurate determination of the organic matter in a soil and the most convenient method is to estimate the oxidizable humus. This is based upon the fact that Hydrogen peroxide removes from the soil that part of the organic material which is easily oxidizable. Since this is the part which is most readily available for plant nutrition, it is perhaps the most important organic constituent so far as vegetation in general is concerned.

Weigh out about 1 gm. of oven-dried soil which has been previously shaken through a 1 mm. sieve. Place it in a tall beaker. Add about 10 c.c. of water and 10 c.c. of 20 vols. Hydrogen peroxide. Warm gently on a water bath, stirring until the evolution of Oxygen ceases. Add another 10 c.c. of Hydrogen peroxide and continue heating for a further five minutes.

Filter, preferably through an asbestos pad, in a Gooch crucible. Wash the residue well with hot water and add the washings to the filtrate. Dry the residue in the crucible at 110° C., cool and weigh. Repeat until a constant weight is obtained.

Transfer the filtrate to a measuring cylinder. Take a measured sample from it and place in a weighed evaporating basin and evaporate to dryness. Ignite the residue, cool and reweigh. The latter weight gives the weight of soluble mineral matter extracted.

The weight of the organic matter oxidized (O) equals the weight of the dry soil (d), less the weight of the residue in the crucible (c), together with the weight of the mineral residue in the filtrate (m). That is:

$$O = d - (c + m)$$

The following are typical results expressed as percentages of the oven-dried soil.

Type of soil	d	c	m	Percentage of organic matter
				per cent
Clay; Lisvane, 1938	10·0	9·5	0·4	3·0
Beechwood; Cranham, 1942 ..	10·0	9·2	0·7	1·0
Limestone meadow; Dinas Powis, 1942	10·0	7·4	0·7	19·0
Peatbog; Margam, 1938	12·2	6·9	0·5	47·5

It must be remembered that the above procedure does not estimate all the organic matter present, especially in soils containing large percentages, where the process of humification is normally slower than in more mineral soils.

Chapter X

LIMNOLOGY; THE ECOLOGY OF PONDS AND STREAMS

A STUDY of water, more particularly of ponds and small streams, very frequently forms a part of survey work. According to the amount of attention which is paid to all the various factors involved, so this branch of the subject may be treated quite simply, or may become very extensive, forming the special branch of science known as Limnology.

While a small pond may form part of a survey area, it is frequently desirable to make a special study of a larger pond or river as an alternative to a land survey. In this case a number of problems have to be investigated, quite apart from the macroflora, the study of which need not be specially referred to here.

In the first place some attempt must be made to investigate the microscopic algae, or plankton, occurring in the water. Secondly, it may be necessary to attempt to discover what fungal organisms are present, since these occur freely in the water either as resting sexual spores or motile zoospores.

The microflora of the pond or stream will be influenced by both the physical and chemical nature of the environment and this involves not only a simple analysis of the water, but also some study of the physical factors affecting the water as well.

The following suggestions, therefore, are given of work along these lines which can be conveniently conducted in the field with the simple apparatus which is included in Box E (see p. 31).

Nets

Nets suitable for plankton work can be bought or made (see Fig. 29). Quite satisfactory nets can be made out of fine bolting silk cut in the form of a cone, the apex of which is open and is firmly attached by means of a tape to the rim of a small bottle or heavy glass tube.

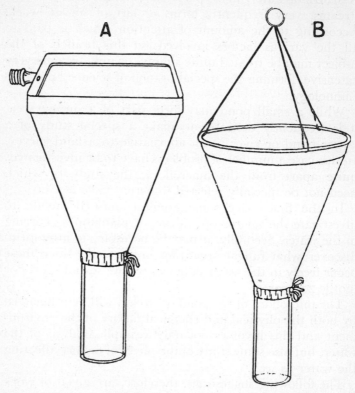

FIG. 29.—Two types of Plankton Nets. A, with a square metal frame and screwed socket for attachment to a rod. B, with metal ring and cotton selvedge, for attachment to a cord for towing. Each is furnished with a detachable collecting tube at the bottom.

Bolting silk is very strong material, used by millers for sifting flour. No. 20 is the best grade for plankton nets, as it has meshes not more than 0·01 ins. across, but it is

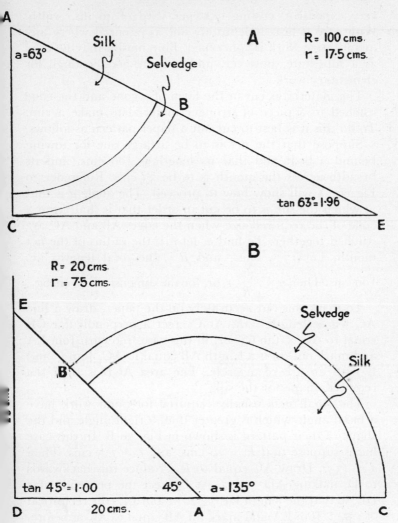

FIG. 30.—Patterns for cutting bolting silk for Plankton Nets. A, where the angle is less than 90°, which is suitable for long towing nets. B, where the angle is greater than 90°, which is more suitable for small nets. For details see text.

very expensive, costing 35s. per yard of 40 ins. width. With care it lasts indefinitely and is essential where any quantitative work is concerned. Fine muslin may be used as a substitute, however, and is quite good enough for elementary work.

The material is cut in the form of a cone and the edge stitched to a piece of stronger material to make a rim. To do this it is best to cut out a paper pattern as follows.

Suppose that the net is to be a large one for towing behind a boat and that its length is 100 cms. and its breadth across the mouth is to be 35 cms. Reference to Fig. 30 A will show how to proceed. The angle a is that at which the silk is to be cut in order that it shall form a cone of the required size when the edges AB and AC are stitched together. To find a, let $r =$ the radius of the net mouth, i.e. $17 \cdot 5$ cms., and $R =$ the total length, i.e. 100 cms. Then $a = \dfrac{360}{R} r$, or, for the dimensions given, $63°$.

To mark this out accurately on the paper, draw a line AC whose length $= R$. At C erect a perpendicular CE equal to AC $\times$ tan $63°$ or, in this case, 196 cms. Join EA and mark out on it a length AB equal to AC. Join C and B with an arc of a circle. The area ACB is then the required shape for the silk.

The small nets usually required for pond work have a basal angle which is greater than a right angle and the shape of their pattern is shown in Fig. 30 B. In this case let us suppose that $R = 20$ cms. and $r = 7 \cdot 5$ cms. Then $a = 135°$. Draw AC equal to R. Produce this backwards to D, making AD $=$ AC. At D erect the perpendicular DE, so that the length DE $= 20 \times$ tan $45°$ (the reciprocal of $135°$). Join EA and mark off AB equal to AC as before.

The net is attached to a frame made of three-eighths inch thick brass rod and is provided with a threaded limb which screws into the top of a handle. If the handle is made of a series of jointed sections a considerably greater reach may be obtained. Alternatively the net may be attached to a ring and drawn across the pond with the aid of a cord.

If a net of this kind is drawn a number of times through the water, the bottle at the end will become filled with a fairly concentrated suspension of the organisms living in the pond water. On the other hand, if a quantitative estimate is to be obtained, a measured volume of water, taken at random over the pond, must be poured through the net. The organisms adhering to the sides of the net must be carefully washed down into the bottle. About 10 litres of water should be used for the initial filtration.

The Rafter Cell

This is a shallow trough (see Fig. 31) holding exactly 1 c.c. in a layer 1 mm. deep. It is made of thin glass strips cemented on an ordinary microscope slide and the

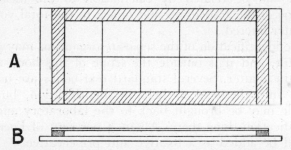

FIG. 31.—A Rafter Cell, made to hold 1 c.c. of solution in a layer 1 mm. deep. A, shows the ruling into square centimeters on the bottom of the trough. B, shows the section with the cover glass in place.

bottom is ruled in 1-centimeter squares. The cell is filled with the suspension from the net, by means of a pipette, and a long cover-glass laid on it carefully, excluding all air bubbles. The slide is then examined under a low power of the microscope for plankton organisms.

Even in qualitative estimations it is desirable to know the relative abundance of the chief organisms. Go over the cell square by square; make a list of, say, half a dozen of the commonest species and record in the field note-book the individuals of each as you come to them, until a total of 100–200 observations has been recorded. The percentage frequency of the species found may then be worked out. As there may be many different species present it is unnecessary and indeed almost impossible for the student to attempt to identify more than a very few of the principal types.

Quantitatively, when a known amount of water has been filtered, the suspension in the bottle is poured off into a measuring cylinder. The Rafter Cell is filled with 1 c.c. and the total number of organisms counted, as well as a relative count being made, as above. This count can be related by calculation to the measured volume of the suspension and hence to the total volume of water filtered.

The identification of the separate organisms may prove difficult, and it is outside the scope of this book to go into the matter. Several standard text-books are usually taken into the field to help in identification, but the sample may be brought back to the laboratory and the identification done there. About 5 per cent of Formalin should be added as a fixative and preservative.

Fungal Traps

Fungal organisms occurring in the water can only be identified by subsequent laboratory study. They belong mostly to the Phycomycetes, especially the Saprolegniaceae. If a few split and boiled hemp seeds are immersed in a sample of the water, which is brought back and placed in a shallow covered dish, growth of these fungi will occur. In about four days the asexual reproductive organs begin to develop. Sexual reproduction begins some days later. Species may be further separated, if desired, by culture on corn meal agar.

Certain fungi will not be obtained in this way, though they are normal inhabitants of pond water. Some may be caught by enclosing fruits; apple, grape or rose hips,

FIG. 32.—Two perforated zinc Cylinder Traps, used to hold a "bait" of rose or other fruits, for aquatic fungi. These traps are attached to lengths of wire and sunk in ponds for several weeks, after which the contents are removed for examination in the laboratory.

in perforated zinc tubes (Fig. 32) closed at one end by a cork. These traps are hung by wires into the water, and the fungi will be found growing on the fruits after a fortnight's immersion.

Even this method will not obtain all kinds, for there remain some which do not attack these baits. They are found chiefly on corticated sticks lying on the bottom of the pond. If such sticks are collected and brought into the laboratory, covered with pond water and left for a few weeks in a very cool place and then brought into the warm laboratory, they will soon become covered with fungal growth, which can be identified by microscopic examination. Not all these fungi can be grown easily on artificial media.

Microscopic Preparations

Where ponds are rich in plankton it is often desirable to obtain permanent preparations which can be studied in more detail in the laboratory than is possible in the field. For this purpose a simple technique is used which will provide satisfactory slides showing the cells in a more or less natural state.

A suitable volume of suspension from the net, the amount depending upon the density of the plankton content, is filtered through a hard surfaced filter paper. The residue is then carefully mixed, with the aid of a fine brush, with a 70 per cent solution of Diethylene glycol in water, containing about 2 per cent of Copper acetate. The mixture may be taken home in a tube, or slides prepared from it immediately. Slides selected for keeping should be carefully packed, as the cover glasses are easily moved. The most satisfactory procedure is to use square cover glasses and seal them immediately with the aid of a mixture of soft wax and resin melted by a wire heated in a spirit lamp.

This mounting liquid will shrink a little by evaporation of water if left unsealed, and in that case any deficiency can then be made good with pure Diethylene glycol containing Copper acetate, before finally sealing the cover glass.

Physical Factors

The physical factors influencing pond waters and the organisms living in them naturally fall under the following main headings: Temperature, Movement, Transparency, and the Nature of the Bottom in so far as it affects the foregoing.

We shall briefly consider each in turn.

1. *Temperature*

The temperature of the water falls rapidly with increasing depth. In order to study this, readings are taken with a self-registering thermometer at known distances apart. The most convenient method is to make a cradle of metal to carry the scale and glass parts of a Six's Maximum and Minimum Thermometer and attach it to a piece of lead. Lower the instrument on a string which has been accurately marked in feet by tying knots in it. The string may be attached to a fishing rod, or to the handle of the plankton net, or to any other long stick that is handy. Read the thermometer, then hold it at a definite depth for a couple of minutes. Draw up the string and read the position of both indices. Any change will show whether there has been a rise or fall in the mercury column, and which index therefore shows the water temperature. Re-set the thermometer with a magnet and lower again to a different depth (see p. 148).

If the temperatures are taken at 1 ft. intervals a graph can be constructed to show the change in temperature of the water over the range of depth tested. Such a graph is intended to show the temperature gradient for the particular pond. More accurate results will be obtained by taking several sets of readings at various parts of the pond and plotting the averages against depth. The best arrangement is to take readings along one or more straight lines across the water, so as to obtain temperature sections of the pond at different positions. In shallow lakes and ponds there is only rarely any marked temperature gradient with depth, since mixing by convection is fairly complete. In such instances it is only necessary to take comparative temperatures at the surface and at the bottom of the pond, measuring the depth with the aid of yard-sticks, and working across from one side of the pond to the other.

The following records may be cited as an example:

OBSERVATIONS TAKEN AT PYSGODLYN MAWR, PETERSTON, GLAMORGAN.
APRIL 29, 1940

Distance from shore (base line) in feet	Surface temperature in °F.	Depth of water in inches	Temperature difference at bottom (− °F)
30	64	9	0·8
60	64	24	2·0
90	65	18	2·0
120	65	24	2·0
150	64	32	3·0
180	63	36	2·5
210	65	32	3·0
240	64	32	2·5
270	64	32	2·5
300	64	34	3·0
330	64	34	3·0
360	65	36	4·0
390	64	30	2·5
420	64	33	2·3
450	63	30	2·0
480	63	26	2·0
510	64	18	1·0

2. *Water Movements*

In flowing water the movements of the water may be approximately estimated by timing the rate of movement of a floating object over a measured distance along the bank. It will be found that the rate of flow in the middle of the stream is faster than near the bank.

In still water the main movements of the water are diurnal and seasonal and are due chiefly to temperature differences and, in a less degree, to wind. In small bodies of water the diurnal movement due to the sinking of the surface water, cooled at night, is enough to keep up a complete mixing. In larger and deeper lakes it needs the cooling effect of winter to bring about complete mixing of surface and bottom water. These movements can be followed only by chemical analysis, especially of the nitrates and phosphates, and they are rather beyond the scope of this book, though their importance is unquestionable. Reference should be made to the books devoted specially to limnology and oceanography for information.

3. *Transparency*

The transparency or turbidity of the water plays an important part in determining the nature of the vegetation. The more turbid the water becomes, the less light passes through it, thereby affecting vegetation growing at the bottom.

Comparative readings of transparency in the field are best made in deep water with a *Secchi Disc*. This instrument consists of a white plate 20 cms. in diameter suspended horizontally by cords. It is lowered into the lake on a graduated rope until it disappears. The depth is recorded. The disc is then raised and the depth at which it reappears is also recorded. The average of these two readings gives the limit of visibility. This method is now used as a standard for comparing the turbidity of water in lakes and in the open sea.

The method may be made more accurate by viewing

H

the disc through a water glass. This consists of a box with a glass bottom, which can be made at home. The box should be about 6 ins. square and 4 ins. deep. Make it up out of quarter-inch wood and make a rabbett or a saw-cut a quarter of an inch from the bottom, all round the inside to take the edge of the glass. Bed the glass firmly in putty and when dry give the whole two good coats of paint to waterproof it.

T	D	T	D	T	D	T	D
7	1,095	20	426	70	138	180	62·4
8	971	22	391	75	130	200	57·4
9	873	24	361	80	122	250	49·1
10	794	26	336	85	116	300	43·2
11	729	28	314	90	110	350	38·8
12	674	30	296	95	105	400	35·4
13	627	35	257	100	100	500	30·9
14	587	40	228	110	93	600	27·7
15	551	45	205	120	86	800	23·4
16	520	50	187	130	81	1,000	20·9
17	493	55	171	140	76	1,500	17·1
18	468	60	158	150	72	2,000	14·8
19	446	65	147	160	68·7	3,000	12·1

T = Turbidity in parts per million, and D = the vanishing depth of the wire in millimetres.

For shallow water in ponds and small lakes the Platinum wire method is more suitable. This method requires a rod with a piece of Platinum wire 1 mm. in diameter inserted 1 in. from the end and projecting at right angles for at least an inch. At the other end of the rod, at a distance of 1·2 metres, a wire ring is placed directly above the platinum wire. Through this the observer looks down in making the readings. The depth at which the Platinum wire becomes invisible is recorded in millimetres. The table on p. 114 gives the United States standard of turbidity by this means, expressed in terms of the equivalent concentration of a standard suspension of precipitated Silica.

4. *Nature of the Bottom*

Sometimes it may be possible, especially in shallow ponds, either to see the bottom or at least to discover its nature by paddling. In deeper water samples may be obtained by dropping a heavy weight into the water with its surface covered with some substance to which the bottom will adhere. Grease or tallow are very suitable. Take a short lead pipe to which a cord has been attached and enclose it in a lump of tallow. Drop this into the water and allow it to sink. Draw it up and examine the material embedded in the tallow. A number of such samplings should be made if the pond is of any appreciable size.

Chemical Factors

The analysis of water samples presents some difficulties, especially if it is carried out in the field. For most of the estimations colorimetric methods are employed and, in order to obtain comparable results, accurately weighed quantities of the necessary reagents are required. Since it is inconvenient to attempt to do this in the field, it is more satisfactory to make use of the "Soloid" tablets put up by Messrs. Burroughs and Wellcome. These have been specially prepared for the analysis of water samples, in connection with pollution of drinking water. The following instructions, therefore, are based on the methods they recommend. The apparatus for making these tests is taken in Box E, together with the other limnological apparatus.

For the analysis of water samples in connection with the behaviour and distribution of phytoplankton, tests must be made for the following: Ammonia, Nitrate, Chloride, Organic Matter, Hardness, Total Dissolved Solids, Dissolved Oxygen and Carbon dioxide, and the Hydrogen ion concentration.

1. *Ammonia*

To 100 c.c. of the water sample contained in a 100 c.c. cylinder, add one 2 c.c. tube of Soloid Nessler Reagent. Break with a glass rod and mix with the water.[1] Note whether there is any colour produced in a few minutes; if yellow, note if faint, distinct or very distinct.

If the colour is very faint, an estimation[2] of the amount of Ammonia present may be obtained in the following way. Place 100 c.c. of the water in a distilling flask capable of holding 200 c.c. Connect with a small receiver immersed in a beaker of water. Heat with a spirit lamp and distil over 20 c.c. of water. When quite cold add one 0·5 c.c. tube of Soloid Nessler Reagent. A yellow tint is produced varying in intensity with the amount of Ammonia present.

A standard for comparison is also made in a 100 c.c. cylinder. Into 50 c.c. of pure distilled water put one Soloid of Ammonium chloride, containing 0·05 mg. of Ammonia, and add one 2 c.c. tube of Soloid of Nessler Reagent. The liquid will turn brownish yellow and each cubic centimeter will correspond to 0·001 mg. of ammonia. Pour this solution into a graduated cylinder until, on looking down through the column and the test solution at the same time, both appear to have the same depth of colour. It is

[1] As the Soloid tubes of Nessler Reagent are expensive, the reagent can be carried in a bottle and measured out with a pipette.

[2] Since this distillation is troublesome to conduct in the field it may be preferable to take a sample of the water to the laboratory and do the test there later.

convenient to use flat-bottomed tubes for this purpose. (Nessler Tubes) or, if available, a colorimeter may be used. If not, place the two tubes half an inch above a piece of white paper and look through them from the top. Measure the number of c.c.s of the standard required to give the same depth of colour as the test sample.

Since each c.c. represents 0·001 mg. of Ammonia, the amount of Ammonia in the pond sample can easily be calculated.

2. *Nitrates or Nitrites*

Take 100 c.c. of water in a cylinder and add one Soloid of Sodium acid sulphate and one Soloid of Metaphenylene diamine sulphate. If a brown colour develops in fifteen minutes, nitrites are present. If not, proceed as follows.

Take another 100 c.c. of the water in a cylinder and dissolve one Soloid of Sodium acid sulphate, add one Soloid of Zinc dust, stir for a short while and allow to stand for five minutes. Filter into another tube and add one Soloid of Metaphenylene diamine sulphate, dissolve and note the colour.

The rapidity with which the colour is produced and the depth of the tint attained are proportional to the amount of nitrate present. With about one-tenth of a part of nitric Nitrogen per 100,000 of water a very pale brown colour appears almost at once, but is still very pale at the end of five minutes. With half a part per 100,000 the liquid is pale brown at the end of one minute and distinctly orange at the end of five minutes. With one part per 100,000 the brown tint appears immediately and the liquid is distinctly orange at the end of five minutes. A concentration of two parts per 100,000 gives a brown tint as the Metaphenylene diamine sulphate is added and turns orange brown in five minutes.

If greater accuracy is required it can be obtained by comparison with distilled water containing a known quantity of nitrate. Take 100 c.c. of distilled water and add one Soloid of Potassium nitrate, containing 0·00144 gm. This corresponds to 0·2 mg. of Nitrogen. By dissolving one or more tabloids a solution may be obtained giving the same colour as the water under analysis. Since one Soloid produces a solution containing 200 mgs. of nitric Nitrogen (i.e. one-fifth part) per 100,000 of water, the nitrate content of the water sample can be calculated easily.

3. *Chlorides*

Measure 100 c.c. of water into a glass cylinder and add one Soloid of Potassium chromate and dissolve. Then add one Soloid of Silver nitrate, dissolve and mix thoroughly. If no red colour is produced add a second tabloid of Silver nitrate and so on until the solution becomes permanently red. This indicates the end of the reaction. Since each Soloid of Silver nitrate corresponds to 2 parts of Chlorine per 100,000 of water the quantity used up in precipitating Silver chloride can be calculated and therefore the amount of chloride present.

As the chloride is most conveniently estimated as Sodium chloride the amount of the latter may be calculated by dividing by 0·6 or multiplying by 1·65.

4. *Hardness*

Hardness is due to carbonates of Calcium and Magnesium dissolved in the water. Together they represent the Total Hardness, which is estimated in degrees, representing parts of carbonate (reckoned as Calcium carbonate) per 100,000 of water.

Measure 100 c.c. of the water into a stoppered cylinder and add one Soloid of soap. Shake vigorously. If a permanent lather is produced the hardness is under 4 degrees. If no lather is formed add another tabloid and shake again. Hardness in degrees is obtained by the following table:

Number of tabloids required to produce Lathering	Hardness in degrees
1	1–4
2	5–9
3	10–14
4	15–19

If hardness is greater than 14 degrees it is better to take half the quantity of the sample, make up with distilled water, and test again.

5. *Organic Matter*

Measure 100 c.c. of the water into a flask and add one Soloid of Sodium acid sulphate. Dissolve one Soloid of Potassium permanganate in 10 c.c. of distilled water and add one c.c. to the flask. Great care must be taken with these tabloids as they are extremely small. Place the flask over a spirit lamp and boil gently for 15 minutes. If the pink colour disappears add another c.c. of Potassium permanganate, boil again, and so on until the colour remains pink.

Each c.c. discoloured in 15 minutes corresponds to 0·1 parts of oxygen absorbed by the organic matter per 100,000 of water.

If the colour becomes brown, or the solution turbid, the addition of a second tablet of Sodium acid sulphate may improve matters.

In cases where it is possible to discriminate between predominantly animal or predominantly vegetable pollution with organic matter, the following methods, due to Dixon and Jenkins, will give more accurate estimations.

Organic Matter chiefly of Plant Origin

Place 100 c.c. of water in a half-litre flask and add 10 c.c. of N/80 Potassium permanganate and 5 c.c. of Sulphuric acid, diluted 1 to 4 with distilled water. Boil quickly for five minutes and continue boiling slowly for ten minutes more. Cool. Add 2 c.c. of 10 per cent Potassium iodide and titrate with N/250 Sodium thiosulphate, using starch solution as the end point indicator.

Organic Matter chiefly of Animal Origin

To 100 c.c. of water in a half-litre flask add 100 c.c. of N/10 Sodium hypochlorite. Boil as above and add 2 c.c. of 10 per cent Potassium iodide and 10 c.c. of concentrated Hydrochloric acid. Titrate with N/40 Sodium thiosulphate, using either starch solution or α-naphthol flavone (0·1 per cent in alcohol) as indicator.

In each case the permanganate or the hypochlorite respectively is reduced by the organic matter and the amount of Iodine liberated from the Potassium iodide is inversely proportional to the amount thus reduced.

Make a blank titration with the same amount of distilled water to which are added the same quantities of reagents. From this it is easy to calculate the weight of permanganate or hypochlorite which has been reduced by the natural water and hence the amount of oxygen taken up by the organic matter per 100,000 parts of water, which is the basis for comparisons.

6. Total Dissolved Solids

This can be determined in the field only with difficulty and is best done subsequently in the laboratory. Take 100 c.c. of the water in a stoppered bottle and remove to the laboratory. Here weigh a dry nickel or silica basin. Pour in the 100 c.c. of water from the bottle, which should then be rinsed out into the basin with distilled water. Evaporate the liquid to dryness over a water-bath, taking care not to allow spluttering or to burn the residue. Cool in a dessicator and re-weigh. The difference in weights gives the total dissolved solids in 100 c.c. of the water, hence the total weight per 100,000 can be calculated.

7. Dissolved Oxygen

The most reliable procedure for field use is the Winkler method. It needs some preparation beforehand but is fairly simple to carry out on the spot.

The following solutions are required:

1. Manganous sulphate, 48 gms. in 100 c.c. of distilled water.
2. Sodium hydroxide 36 gms. and Potassium iodide 10 gms. in 100 c.c. of distilled water.
3. Concentrated Sulphuric or Hydrochloric acid.
4. Sodium thiosulphate N/100. This solution does not keep indefinitely. It keeps better if 5 c.c. of chloroform and 1·5 gms. of Sodium carbonate are added to each litre.
5. Starch solution as an indicator for the titration of Iodine.

Collect the water sample in a well-stoppered bottle of about 250 c.c. capacity. The bottle must be completely filled and the water must have as little contact with air as possible.

With a pipette run in 1 c.c. of solution 1 and 1 c.c. of solution 2 to the bottom of the bottle. Replace the stopper without any air bubbles and mix by rapidly inverting and rotating the bottle. In the same way add 2 c.c. of the acid and again mix. When all the precipitate formed has dissolved, transfer 100 c.c. of the solution to a small flask and titrate with solution 4, using solution 5 to indicate when all the liberated Iodine has combined with the thiosulphate.

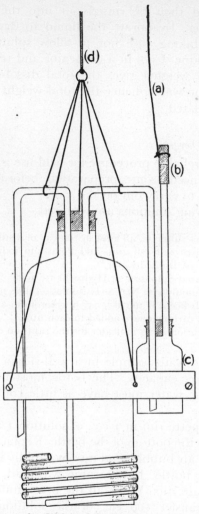

FIG. 33.—Diagram of a Weighted Bottle, showing the arrangement for obtaining a sample of water from a required depth. The rubber tube (*b*) is blocked with a piece of glass rod. The rubber can easily be pulled off by means of the free string (*a*). The small bottle (*c*) takes the overflow of water which has been in contact with the air in the main bottle. (*d*) the main supports for lowering the bottle.

If X is the number of c.c. of N/100 thiosulphate used in the titration, then the dissolved Oxygen in parts per 100,000 is 0·378X and in c.c. per litre it is 2·645X.

It was pointed out above that a sample of water for the estimation of the Oxygen content must be collected with as little contact as possible with the air. This applies with particular force to samples taken from any depth below the surface. For this purpose a special bottle should be prepared, which can also be used to collect depth samples for other purposes.

Remove the stopper from a 250 c.c. bottle and fit it with a double-bored cork. Through one hole passes a tube bent twice at right angles, with a long limb going down to the bottom of the bottle and a shorter limb lying against the side of the bottle, pointing downwards. Through the other hole passes a similar tube, whose short limb is flush with the underside of the cork and whose long limb passes to the bottom of a second small bottle which is strapped to the side of the large one (see Fig. 33). The small bottle has a straight outlet tube, directed upwards and closed by a clipped rubber tube, fitting not too tightly. The whole can be weighted with a coil of lead "compo" pipe, round the base. The apparatus is sunk to the required depth and the rubber tube is then pulled off by means of a fine cord which has been tied to it previously. The air escapes via the small bottle and both bottles fill, but the first water to enter, which makes contact with the air, passes over into the small bottle and is rejected, while the large bottle contains a good sample. As soon as the bottles have been raised to the surface the cork is carefully removed and replaced by the glass stopper. The sample is then ready for analysis. If the sample is to be kept for any length of time before use than add 2·5 gms. of Sodium formate as a preservative.

8. *Dissolved Carbon Dioxide*

Take 100 c.c. of the water and add 1 c.c. of Phenolph-thalein, prepared as 1 : 1,000 of Phenolphthalein in absolute Alcohol. Titrate with N/50 Sodium hydroxide until a pale pink colour, which lasts for at least three minutes, is produced.

The Sodium hydroxide must be standardized in advance against standard Oxalic acid, working to the same pink colour of the indicator as a standard of comparison.

One mg. of Carbon dioxide is equivalent to 0·9091 mg. of Sodium hydroxide. This method can only be used when the total content of dissolved electrolytes does not exceed 1 gm. per litre. Free Carbon dioxide will only occur in acid waters.

In neutral or alkaline waters Carbon dioxide must be estimated as bicarbonate by the following method, using 1 : 1,000 Methyl Orange in distilled water as indi-cator.

Take 50 c.c. of the water, add a few drops of Methyl Orange and place in a cylinder provided with a cork. Titrate with N/25 Hydrochloric acid, removing and replacing the cork each time the acid is added, to keep in any free Carbon dioxide. Standardize the Hydrochloric acid against Sodium hydroxide, using Phenolphthalein as indicator according to the method given above.

One mg. of bicarbonate Carbon dioxide is equivalent to 0·83 mg. of Hydrochloric acid.

9. *Hydrogenion Concentration*

The pH of the water sample requires more careful estimation than soil extracts, especially if seasonal ob-servations are to be made, as the amount of variation is usually small. A few c.c.s of the water should be collected in a small vessel which has been previously well washed out with the same pond water. Water from the surface or from a measured depth should be taken. In the latter case the weighted bottle should be used which has been described under the estimation of dissolved Oxygen (see p. 123). Remove the sample from the bottom of the bottle with a pipette. From this sample take a drop in the measured capillary tube and proceed as described for the use of the Capillator with soils (see p. 94).

Readings may be made at different depths, or samples

TABLE OF pH MEASUREMENTS TAKEN MONTHLY FROM THE SURFACE WATER OF A LARGE POND, PWLL DIWAELOD, CASTLETON, MONMOUTHSHIRE, DURING 1941

Month	pH	Month	pH
January ..	6·4	July	7·0
February ..	6·4	August ..	7·0
March ..	6·9	September ..	6·9
April ..	7·0	October ..	6·8
May ..	7·0	November ..	6·8
June ..	7·0	December ..	6·6

may be taken at surface level at different time intervals to study any seasonal changes.

The table above shows the change in pH of the surface water of a pond observed monthly throughout the year. It will be seen that the pH fell slowly through the autumn so that the water was most acid in January and February and remained more or less at a constant pH from April to August.

This pool is deep all the year round and is situated in a meadow, lying on Old Red Sandstone and Marls underlying glacial drift of gravels and sands. The pond is surrounded by trees and there is no growth of plants on the surface of the water.

The following analyses of waters from Glamorgan may be quoted as examples showing the variation which occurs in the composition of the water under different environmental conditions. Pysgodlyn Mawr is a large, shallow lake on gravel, surrounded by peat bog. Kenfig Pool is a coastal lake of fresh water surrounded by sand-dunes, while the pond at Dinas Powis is typical of agricultural land on limestone.

Substance tested	Pysgodlyn Mawr in parts per 100,000	Kenfig Pool in parts per 100,000	Dinas Powis pond in parts per 100,000
Carbon dioxide content	14·5	9·6	13·2
Total dissolved solids	41	10	20
Hydrogen ion concentration at surface ..	6·0	7·0	7·2
Hardness	1–4 degrees	4–9 degrees	14–19 degrees
Nitrate content ..	nil	nil	0·1
Ammonia content ..	nil	nil	nil
Chloride content ..	10	10	13·3
Total organic matter	1·9	0·2	0·6

Chapter XI

CLIMATIC AND PHYSIOLOGICAL FACTORS

THE investigation of climatic and physiological factors in relation to ecological surveys presents very considerable difficulties. Many of the problems can only be solved by continued observation over long periods. Climatic factors must for the most part be studied throughout the whole year, and an isolated set of readings made on a single occasion is of comparatively little value.

The main climatic factors naturally fall into the following five headings: rainfall, atmospheric humidity, sunshine, temperature and wind. While it is true that in the main these factors must be regarded as more or less uniform over the area of survey, a more critical study will reveal that it is not strictly true. In recent times we have come to appreciate the importance of a micro-climate. A rock, or even a stone, may play quite an important part in influencing the environment of the plants in its immediate neighbourhood. On sloping ground a stone may hold up water, thus making the soil on the higher side of a slope appreciably wetter than the lower. It may protect the plants by breaking the force of the prevailing wind and in this way may also increase the temperature on the protected side. Such points are recognized and used in connection with modern rock gardening, where the stones are placed to serve a purpose and are not merely dotted about at random. It is obvious, moreover, that plants themselves may perform much the same function in respect to their neighbours. A dense evergreen bush will protect delicate plants, a practice frequently employed in gardens to protect tender plants from frost and cold winds. Other examples will readily occur to the thoughtful investigator.

It is, therefore, necessary to consider these factors in detail in connection with a study of plant ecology, though it will be realized that some are more easy to study than others.

1. *Rainfall*

Estimates of rainfall are made at all meterological stations and many schools record the daily rainfall. A rain gauge is used for the purpose. It consists of a circular copper pot 5 ins. in diameter (see Fig. 34), the inside

Fig. 34.—A good pattern of Rain Gauge, showing the protecting rim which prevents rain being blown out of the funnel. The diameter of the funnel is 5 in. The water is collected and measured in the cylinder, which is graduated to read directly in fractions of an inch.

being formed into a funnel which connects with a measuring cylinder in which the quantity of water collected is measured. Generally the cylinder is graduated so that it gives the amount of water collected in the gauge in terms of hundredths of an inch on the surface area. If a rain gauge is not available on the area itself records should be obtained from the nearest permanent station, the situation of which can be obtained from the Meterological Office.

Local records are of considerable value and may be expressed in the form of a graph, where the depth of water is recorded on the vertical scale and the time intervals on the horizontal. The time interval selected is 24 hours, from 9 a.m. to 9 a.m.

2. *Humidity*

The relative humidity of the air is a factor which should be recorded during the period of the survey, for it is

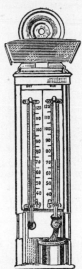

FIG. 35.—A typical Wet and Dry Bulb Thermometer. The wet bulb is kept moist by a wrapping of absorbent material, which dips into a reservoir of water.

closely related to the rate of transpiration of the vegetation. Relative humidity means the percentage saturation of the air with water vapour.

I

Direct readings of the humidity of the air may be made most accurately with a wet and dry bulb thermo-meter (Fig. 35). This instrument consists of two identical thermometers set up side by side on a board. The bulb of one is left exposed while that of the other is covered with a piece of absorbent material connected with a small water reservoir. The instrument is placed in a shady place,[1] not exposed to the direct rays of the sun,

CONVERSION OF READINGS OF WET AND DRY BULB THERMOMETER INTO RELATIVE HUMIDITIES

Temp. °C.	Depression of wet bulb thermometer in degrees Centigrade									
	1·0	2·0	3·0	4·0	5·0	6·0	7·0	8·0	9·0	10·0
0	81	64	46	29	13					
+ 3	84	69	54	40	25	12				
+ 6	87	73	60	47	35	23	11			
+ 9	88	76	65	53	42	32	22	12		
+ 12	89	78	68	58	48	38	30	21	12	4
+ 15	90	80	71	62	53	44	36	28	20	12
+ 18	90	82	73	65	57	49	42	35	27	20
+ 21	91	83	75	67	60	53	46	39	32	26
+ 24	92	85	77	70	63	56	49	43	37	31
+ 27	93	86	79	72	65	59	53	47	41	36
+ 30	93	86	79	73	67	61	55	50	44	39

and after allowing time for the instrument to become adjusted to the environment the temperatures on both thermometers are read. The difference between the two gives a value which can be used, by reference to the above tables, to determine the relative humidity at the temperature given by the dry bulb.

[1] A quicker and more accurate reading is obtained, by attaching a piece of thick cord to the mounting board and swinging the whole instru-ment rapidly round the head.

Alternatively the humidity may be determined directly with a hair hygrometer. This instrument depends on the lengthening of a hair according to the amount of moisture in the atmosphere. One end of the hair is fixed, the other is connected with a pointer which moves over a graduated scale which gives direct readings of the percentage of

FIG. 36.—Paper-strip Hygrometer. A suitable type for field work.

humidity in the atmosphere. When new these instruments are highly satisfactory, but they need checking at frequent intervals if accuracy is required.

Finally, there is the paper-strip hygrometer (Fig. 36), which is the most convenient type for use in the field. A small strip of sensitive material is arranged to activate a pointer which moves over a scale reading from 0 to 100 per cent saturation. If used as a stand instrument it may be supported on its own foot, in which case it takes about 10 minutes to come to rest and give a true reading of the percentage humidity. Alternatively the instrument may be fitted to a metal folding rod attached to a handle, which allows the instrument to be whirled around the head. Since the side of the hygrometer is perforated the sensitive strip is thereby brought into moving contact with the atmosphere and immediate readings are obtained. Care must be taken to protect the instrument from rain and it should not be placed in the direct rays of the sun.

3. *Atmometry*

The humidity of the atmosphere is directly related to the rate of evaporation of water, either free or from a porous surface. The bulb type of atmometer was perfected by Livingston, and this pattern is well adapted to field observations. Essentially it consists of a porous porcelain bulb of known surface area, which is filled with distilled water and connected with a capillary tube into which a bubble can be introduced, that is to say, a potometer in which the bulb replaces the plant.

A spherical bulb is the best form for long continued readings, as it is uniformly affected by the movements of the sun across the sky, but for short readings an ordinary porous filtering candle, closed at one end, is perfectly satisfactory and is less bulky. It has the additional advantage of offering a greater evaporating surface per unit volume than does a sphere.

The type of apparatus which we have found most convenient is shown in Fig. 37. One important factor which had to be considered in its design is temperature. If a large volume of water is used as a supply for the apparatus, the rate of movement of the air bubble will be largely influenced by changes of temperature. In extreme cases this may be sufficient to cause the bubble to move backwards, indicating that the increase in

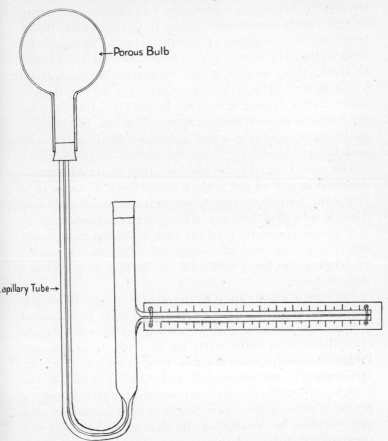

FIG. 37.—Diagram of a convenient type of Atmometer, provided with a porous Livingston bulb and capillary tube, with a water reservoir and a side tube provided with a scale. This pattern is suitable for carrying up trees and other difficult places.

volume due to the rise in temperature is greater than the decrease in volume due to evaporation. Wherever possible, therefore, the apparatus should be used in the shade, and in any case sufficient time must be allowed for the apparatus to acquire a temporary equilibrium, as shown by the bubble movement coming to a steady rate, before recording the reading. The apparatus is mounted for work in the field by attaching a board, shaped as a handle, to the U-tube, so that it can be carried about without touching the glass. Alternatively it can be fixed in a retort clamp. During transit in the field it is best to cover the bulb to prevent evaporation. This can be done by enclosing it in a 4-oz. tobacco tin provided with a split cork bung.

The time taken for the bubble of air to travel along a standard length of the scale is timed with a stop-watch. When the bubble escapes from the capillary it rises into the side tube and does not affect future readings.

A simple comparison of the rate of evaporation under different environmental conditions can thus be made, by relating the times taken for the bubble to travel the same distance.

The rate of evaporation from the porous bulb should be calibrated in terms of the evaporation from unit area of a free-water surface. Fill a vertical-sided vessel, e.g. a crystallizing dish, with water up to 1 cm. from the rim. Weigh the whole. Place the dish in a position close to the atmometer to be calibrated and take readings of the bubble movement in the instrument at intervals over a period of some hours. These readings must be transformed into volumes by measuring the diameter of the bore of the capillary tube and the length over which the bubble has been timed. Measure the area of the porous surface and calculate the average rate of water evaporation per unit area per unit time, e.g. per square decimeter per hour.

Re-weigh the water vessel at the end of the period and calculate to similar terms the water lost by direct evaporation from the surface. The ratio of the two rates is the Efficiency Factor of the bulb, by the use of which its readings may be reduced to terms of evaporation from an equivalent free water surface. In this way a number of atmometers can all be brought into comparison with each other.

An interesting type of evaporimeter has been suggested by Bates, of which, however, the authors have had no practical experience. It might be worth experimenting with it in studies of woodland conditions. A sheet of linen is enclosed between an upper plate of blackened metal and a lower plate of perforated zinc. The end of the linen dips into water in a graduated vessel to keep it moist by capillarity. The blackened upper surface absorbs radiant energy and the lower perforated plate simulates a stomatal leaf-surface. Evaporation from this instrument is said to follow the transpiration rates of young Conifers very closely.

4. *Transpiration*

There is a direct relationship between the rate of evaporation, as measured by an atmometer, and the rate of transpiration of leaves. This may be best measured in the field by the Cobalt chloride method. This method is based upon the fact that Cobalt chloride is deep blue when quite dry but pale pink when wet. Make up a 5 per cent solution of Cobalt chloride in water and dip a number of strips of filter paper in the solution. They will appear pale pink in colour. Put them in an oven to dry, when they will change to a deep intense blue. These papers may be cut into one inch squares and stored in well-stoppered bottles, in the bottom of which is some Calcium chloride, separated from the paper by a sheet of perforated zinc cut to size.

The water vapour given out by the stomata is enough

to change the colour of the paper, so a measure of the rate of transpiration may be found by timing the change from the original blue, either to a pale blue or from pale blue to the appearance of the pink. The latter is preferable, as it obviates errors due to damp leaves.

Take an ordinary microscope slide and stick three strips of coloured paper about one-quarter of an inch apart on the centre of the slide. One should be the same colour as the dry Cobalt chloride paper, one a very pale blue and the third pale pink. Coloured paper may be used or suitable paper may be obtained by staining with Methylene blue and Eosine. An arbitrary standard of colour may be chosen, but must be adhered to throughout the whole series of observations.

The experiment is set up as shown in Fig. 38. Two slides are used, one above and one below the leaf, and between the leaf and the slide, on either surface, a square of Cobalt chloride paper is placed. The leaf is left on the plant during the experiment, and tests are made on various leaves and with different individuals, growing in different positions, so that measures of the actual transpiration may be obtained under a variety of external circumstances. Observe, with a stop-watch, the time taken for both sides of the leaf to turn the Cobalt chloride paper to the pale pink colour. It will be realized that only perfectly dry foliage can be used. In fact work on atmometry and transpiration only yield the best results on fine, dry days. It is desirable that an atmometer observation should be taken in close proximity to the transpiration tests.

Sometimes several different species of plant may be tested in the same plant community. Alternatively only one species is used, but the selected plants are those growing in different environments. Records may be made of leaves at different heights, as in the case of a tree. In this case the atmometer reading must be made at the same height up the tree. Remember that the rate of evaporation high up *inside* the crown of a tree is usually

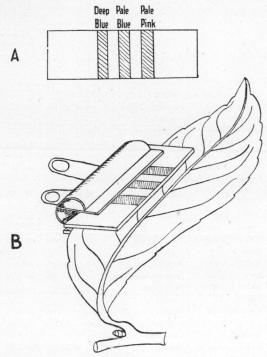

Fig. 38.—Cobalt Chloride Method. A, microscope slide with three standard tint strips attached. B, method of attaching the slides and clip to the leaf still on the plant.

much less than at the same height *outside* the tree, and the same is true of the rate of transpiration.

The foregoing results, taken with the apparatus described, may be quoted as fairly typical. In this case two sets of students were working together, the one doing the atmometry observations, the other the transpiration,

RELATION OF ATMOMETRY AND TRANSPIRATION RECORDED NEAR
BOSSINGTON, SOMERSET, 1939

	Species used: *Pteridium aquilinum*				
	Habitat	Conditions	Evaporation rates c.c./met.^{2}/min.	Relative transpiration rates $1/T$ (min.)	
				Upper	Lower
L.* {	Wood	Deep shade, still	$1 \cdot 16 \pm 0 \cdot 12$	$0 \cdot 051$	$0 \cdot 078$
	Roadside	Shade, slight wind	$4 \cdot 01 \pm 0 \cdot 48$	$0 \cdot 25$	$1 \cdot 0$
Ex. {	Under beech; hillside	Shade, slight wind	$6 \cdot 63 \pm 0 \cdot 85$	$0 \cdot 20$	$0 \cdot 3$
	Open hillside	Sunny, windy	$5 \cdot 05 \pm 0 \cdot 21$	$0 \cdot 20$	$0 \cdot 3$
S. {	Under bracken; hillside	Shade, still	$6 \cdot 08 \pm 0 \cdot 39$	$0 \cdot 20$	$0 \cdot 5$
	Under *Castanea*; wood	Shade, still	$3 \cdot 23 \pm 0 \cdot 01$	$0 \cdot 25$	$0 \cdot 5$
	Under *Castanea*; wood	Shade, still	$3 \cdot 23 \pm 0 \cdot 01$	$0 \cdot 105$	$0 \cdot 167$
	Under *Acer*; wood	Shade, still	$2 \cdot 94 \pm 0 \cdot 15$	$0 \cdot 062$	$0 \cdot 182$
	Under *Quercus*; hillside	Sun flecks, wind	$1 \cdot 28 \pm 0 \cdot 08$	$0 \cdot 092$	$0 \cdot 297$
	Open hillside	Sunny, windy	$5 \cdot 12 \pm 0 \cdot 48$	$0 \cdot 22$	$0 \cdot 40$
	Open hillside	Sunny, windy	$5 \cdot 12 \pm 0 \cdot 48$	$0 \cdot 182$	$0 \cdot 40$
	Under *Quercus*; hillside	Sun flecks, wind	$9 \cdot 16 \pm 0 \cdot 33$	$0 \cdot 20$	$0 \cdot 364$
	Open hillside	Sunny, windy	$3 \cdot 70 \pm 0 \cdot 19$	$0 \cdot 33$	$1 \cdot 00$
	In *Quercus*; thicket	Sun flecks, windy	$3 \cdot 20 \pm 0 \cdot 80$	$0 \cdot 087$	$0 \cdot 33$
	Under wall; edge of wood	Shade, still	$2 \cdot 45 \pm 0 \cdot 07$	$0 \cdot 20$	$0 \cdot 40$
	Species used: *Fagus sylvatica*				
L.	Clearing in wood	Sunny, still	$1 \cdot 37 \pm 0 \cdot 16$	$0 \cdot 50$	?
Ex. {	In tree, 3 ft. up	Sun flecks	$6 \cdot 53 \pm 0 \cdot 46$	$0 \cdot 286$	$4 \cdot 27$
	In tree, 30 ft. up	Shade	$6 \cdot 53 \pm 0 \cdot 46$	$0 \cdot 143$	?
	Under tree	Shade, windy	$6 \cdot 63 \pm 0 \cdot 85$	$0 \cdot 311$	$4 \cdot 29$
	Open moor	Full sun, windy	$2 \cdot 39 \pm 0 \cdot 89$	$0 \cdot 313$	$7 \cdot 84$

* L. = West Luccombe Wood. Ex. = Simonsbath, Exmoor. S. = Selworthy Wood.

RELATION OF ATMOMETRY AND TRANSPIRATION RECORDED NEAR
BOSSINGTON, SOMERSET, 1939—*continued*

	Habitat	Conditions	Evaporation rates c.cs./met.^{2}/min.	Relative transpiration rates 1/T (min.)	
				Upper	Lower
L.	Roadside	Sunny, slight wind	4.63 ± 0.56	0.25	2.75
Ex.	Shelter of hedge; moor	Shade, still	5.23 ± 0.41	0.50	?
S.	Under *Castanea*; wood	Shade, still	3.23 ± 0.01	1.00	8.55
	Open hillside	Sunny, windy	5.12 ± 0.48	1.00	?
	Under *Quercus*; open hill	Sun flecks, very windy	9.16 ± 0.33	0.667	4.66
	Under wall; edge of wood	Shade, still	2.45 ± 0.07	0.667	3.67

Species used: *Urtica dioica*

L = West Luccombe Wood, Ex. = Simonsbath, Exmoor, S. = Selworthy
Wood.

This example shows the method of expressing the results
when finally worked up in the evening. Where possible,
the mean of half a dozen observations in each locality
should be given. The transpiration rates are given as
the reciprocals of the time in minutes for the colour-
change to take place.

Observations of this kind should always be made
where possible on the actual survey area and should be
regarded as an essential part of the work, though more
striking results are generally obtained in woodlands and
on open moors than in meadows and marshes.

5. *Light*

The intensity of illumination may play an important
part in determining the nature of the prevalent plant

life, especially in a wood. Shade from trees, if sufficiently heavy, may almost, if not quite, suppress any ground flora, and in most woodlands the intensity of illumination is a limiting factor in distribution so far as the ground flora is concerned, though among shallow-rooted trees, such as Birch or Pine, root competition may be the decisive factor, and this should be borne in mind.

The measurement of the light in absolute units, i.e. metre-candles or "lux," is difficult and involves the use of a photo-electric cell, which is generally outside the range of most field classes. Certain photo-electric light meters, used in photography, can be employed, however, and their prices are more reasonable. Unless the instrument is calibrated beforehand such readings are only comparative, but even so they are of value to compare the intensity of light in different parts of a woodland survey.

The method employed is to hold the instrument up at an angle of 45° facing a piece of white paper held horizontally above the ground vegetation and to observe the reading when the instrument is set for a particular standard of stop and plate selected for average sensitiveness. Using a purely arbitrary setting of this kind quite interesting comparative readings can be made. It must be pointed out, however, that profound differences will be produced according to whether the sun is obscured by cloud or not. Theoretically, all comparative readings should be made simultaneously, but, as this is usually impracticable, care and judgment must be used to select conditions which are as uniform as possible.

A reading of light intensity should be made in the centre of each square of the grid as far as possible under uniform cloud conditions. The results are best expressed as the reciprocals of the indicated photographic exposures, in seconds, or they may be reckoned as fractions of the maximum observed intensity.

Exposure meters of the paper-blackening type may also be employed to determine intensity of light. Such papers change colour from pale yellow to black on exposure to light. Generally they are matched against a semicircular standard grey-black paper, and the time taken to change from the yellow to the standard tint is taken with a stop watch. This time is used as a basis of comparison, or the instrument may be calibrated by timing the exposure against a standard illuminator or by exposing the meter to daylight, which is simultaneously measured in lux with a photo-electric meter.

The following comparative figures have been published by Bracher for the Watkin Bee Meter, which is a very convenient form of instrument for the purpose. The same ratios, of course, equally apply to other paper actinometers.

Boysen Jensen has described a method using Rhodamine B paper for estimating and comparing the intensity of illumination. The advantage of this paper is that it is specially sensitive to yellow light and therefore gives a

Time in secs.	Relative intensity	Time in secs.	Relative Intensity
1	60	13	4·6
2	30	14	4·3
3	20	15	4·0
4	15	20	3·0
5	12	25	2·4
6	10	30	2·0
7	8·6	35	1·7
8	7·5	40	1·5
9	6·6	45	1·3
10	6·0	50	1·2
11	5·4	55	1·1
12	5·0	60	1·0

measure of illumination which reflects the conditions affecting photosynthesis better than the ordinary sensitive paper, which is chiefly affected by the shorter wave lengths of the spectrum.

The paper must be specially prepared. Cut up good, fairly heavy, unglazed writing paper, into pieces about 3 ins. square, and soak for five minutes in a 60 per cent solution of Potassium bromide. Then hang them up vertically to dry. Float the sheets carefully on a bath of 12 per cent Silver nitrate solution, in the dark, for 2 minutes. Transfer to water and wash thoroughly for 24 hours also in the dark.

Finally soak for five minutes in the following bath:

Sodium nitrite	..	..	6 gms.
Water ..	..	..	.. 200 c.c.
Rhodamine B solution		..	12 c.c.

The Rhodamine B solution is made by dissolving 1 gm. in 200 c.c. of absolute Alcohol. Hang up the paper to dry. This paper is red and goes black in the presence of light.

The times given above are for baths at about 20° C. Colder baths require longer periods of soaking. Test clippings from the prepared sheets, exposed side by side, and reject any sheets which depart markedly from the average in colour or sensitivity.

A standard tint for comparison is made in the same way, but the Sodium nitrite is omitted. Such paper blackens very slowly in light and the colour formed is practically permanent.

A convenient form of holder may be made as follows, (see Fig. 39). Take an ordinary shallow round tin, such as is used for vaseline or typewriter ribbons. Select one whose lid moves easily though firmly on the bottom. In the bottom fit a pad of felt and wood sufficiently thick for

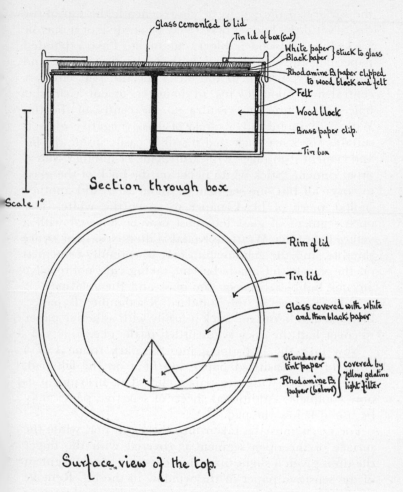

Glass cemented to lid
Tin lid of box (cut)
White paper } stuck to glass
Black paper }
Rhodamine B paper clipped to wood block and felt
Felt
Wood block
Brass paper clip.
Tin box

Section through box

Scale 1"

Rim of lid
Tin lid
Glass covered with white and then black paper
Standard tint paper } covered by yellow gelatine light filter
Rhodamine B. paper (below) }

Surface view of the top.

FIG. 39.—Rhodamine B. Light Meter. Section through box showing the way in which the various parts are arranged and, below, a plan of the top to show the segment cut out of the paper and the yellow screen through which the standard and sensitive papers are viewed.

the surface of the felt to lie close beneath the top of the tin. Several thicknesses of felt may be needed, or it may be backed by a wooden block cut to fit. Pass a pointed paper-clip through the pad, with the head upwards, and clip over. Now turn to the lid of the tin. Cut either a central hole about $1\frac{1}{2}$ ins. in diameter, or cut a circular hole $\frac{3}{8}$ in. diameter eccentric to the centre of the tin. With a diamond cut a circle of glass (negative glass is suitable) of a size such that it will fit easily inside the lid and cement it in place with Murrayite, or another water-proof cement. Stick white paper on the back of the glass to cover all but one segment of about 20°. Add another similar piece of black paper to cover the white. The open segment of glass left must now be covered with a yellow light filter. Wratten (Kodak) Filters Nos. 12 or 15 are suitable, and the gelatine film may be carefully cemented on the glass with Canada balsam, taking care not to leave any air bubbles between the glass and the gelatine.

Take a piece of the standard Rhodamine B paper, blackened in light, and stick it firmly with adhesive paper to cover half the open segment left in the glass.

Now take the instrument into the dark room. Cut a circle of the sensitized paper and fit it on the felt pad, making it fast with the paper clip. The instrument is now complete. Additional sheets of sensitive paper may be stored below the pad.

For use it must be taken into the light, and while the surface of the open segment is covered with the finger the tin is given a slight turn, thus exposing a new segment of the sensitive paper at the window in the lid. Remove the finger and take the time required to change the paper to the same tint as the standard. Successive readings are made by turning either the bottom or the lid the width of one segment each time. Incidentally, precisely the same method is used with the Watkin Bee Meter, and by altering the standard paper of one such meter it can equally well be used with the Rhodamine B paper. Since the standard paper for these exposure meters is now very difficult to obtain, Rhodamine B paper might be suitably substituted, though of course it does not follow that for photographic purposes the factors would be the same.

If possible the Rhodamine B paper should be standardized to find out the time taken to colour the paper to the standard tint when exposed to light of known intensity. Since the paper is relatively slow in action, tests with artificial light are not very accurate, and if possible a photoelectric light meter should be employed as a standard for comparison. In such cases a graph can be constructed to show the actual light intensity in lux which corresponds to a particular exposure.

Unfortunately it is difficult to obtain a truly standard paper by the method described, and separate batches of sensitized paper may vary considerably among themselves, so that it is impossible to give a conversion table which would be of any general value.

In certain cases, where exposure to sunlight is specially intense, as, for example, on the sea shore or on high mountains, it may be desirable to get some idea of the ultra-violet component in the sunlight. This can be done very simply by covering the sensitive paper with a gelatine screen containing Aesculine (Wratten (Kodak) Filter No. 2), which is colourless, but cuts out the ultra-violet sharply. The difference between exposures made with and without the screen give a measure of the relative intensity of the ultra-violet component in different situations and at different times.

The curves shown in Fig. 40 give an example of such readings made throughout a cloudless summer day near the sea coast.

K

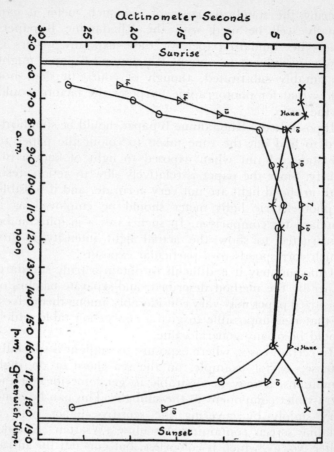

Fig. 40.—Curves showing the typical variation of full sunlight and of the ultra-violet component, measured in bright sunlight on open hill top, 200 ft. above sea and with cloudless sky. Overton, Gower, Glamorgan, Sept. 1928. The numbers above the points give the dates when the observations were made.

Δ = full normal light (W).
O = light measured through an aesculine screen (A).
X = $1/A - W \times 10$, or the reciprocals of the differences between the two factors.

6. *Temperature*

Studies in changes of temperature can be carried out in the field and provide interesting data.

The difference in temperature between plants and their surroundings is often striking. For example, comparative tests can be made showing the differences in temperature in the open air and in the centre of tufts of low, dense growing plants. The following observations may be quoted as an example.

COMPARATIVE READINGS OF TEMPERATURE TAKEN AT SIMONSBATH,
DEVON, 1939

Species	Temperature in the sun: 1 ft. above earth	Temperature within the plant tuft	Difference in temperature
	°C.	°C.	°C.
Juncus effusus	24·2	30·2	6·0
Agrostis stolonifera ..	22·0	28·0	6·0
Sphagnum sp.	14·0	19·5	5·5
Sphagnum sp.	15·5	19·0	3·5
Narthecium ossifragum	14·7	16·5	1·8
Eriophorum angustifolium	15·3	16·5	1·2
Molinia coerulea	15·8	17·0	1·2
Juncus squarrosus ..	17 5	18·0	0·5
Sphagnum sp. (dead)	16·0	16·0	0·0
Molinia coerulea (dead)	16·0	16·5	0·5

In this experiment paired thermometers were used. One was held one foot above the ground, the other buried in the centre of the plant. The differences are striking, whether they are due to respiration or to absorption from sunlight or both. Similar effects may be obtained by comparing the sunshine and shade influence on the temperature of the ground vegetation in a wood. There are many similar experiments which can be tried out by students, in particular situations.

For such comparisons sensitive thermometers are required and a pair reading to 0·1° C. should be used.

Temperatures at different depths in water can be taken

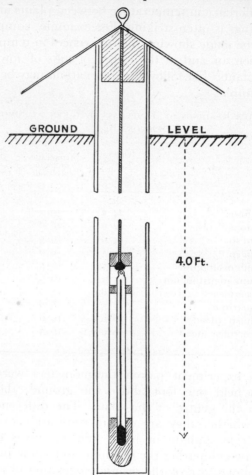

GROUND LEVEL

4.0 Ft.

FIG. 41.—Diagram showing the construction of the housing for a Soil Thermometer arranged to work 4 ft. below the ground.

by attaching a thermometer to a length of string measured in feet. A weight may be attached to the string to act as a sinker. Provided that the depth is not great, fairly accurate measurements can be obtained by coating the thermometer bulb with paraffin wax as a partial insulator, and then lowering it to the desired depth on a string. After leaving it there long enough to become adjusted it is pulled up and read immediately. Self-registering thermometers are, of course, more accurate.

Temperatures in the soil are frequently taken, but rarely on the survey area. Such observations are better made in more permanent quarters near the laboratory. Generally the soil temperatures are registered at 1 ft. and 4 ft. An iron tube about 2 ins. in diameter is sunk into the ground to a little more than the desired depth (see Fig. 41). About 6 ins. or a foot of tube is left exposed at the top. A thermometer is enclosed in a glass tube about 1 in. in diameter, which is well corked at the top and filled at the bottom with paraffin wax. This tube serves to protect the thermometer and also acts as a partial heat insulator when the instrument is drawn up for inspection. A chain is attached which ends in a cork, fitting into the top of the metal tube. A conical roof of zinc is provided to prevent water getting into the tube. Such thermometers will give a reading of the soil temperature and show the degree to which the fluctuations in atmospheric temperature affect the soil. The table on page 150 gives comparative readings at 4 ft. below and 4 ft. above ground level during the weeks nearest midsummer and midwinter. In both cases the readings were taken at 11.0 a.m.

Temperatures of surface soil or a few inches below the surface can often be obtained simply by carefully thrusting a chemical thermometer of the solid glass type into the soil to the required depth.

Thermographs are occasionally available. These instruments magnify by levers the expansion of a strip of metal. The end of the lever is provided with a pen which moves over the surface of a drum, covered with a paper scale. The drum is arranged to rotate once in a week. Thus a complete history of the changes in temperature day by

TABLE OF COMPARATIVE TEMPERATURES FOR MIDSUMMER AND MID-WINTER, TAKEN AT 11.0 A.M. DULWICH COLLEGE, 1914[1]

Date	Temperature 4 ft. above ground	Temperature 4 ft. below ground	Difference in temperature
	°F.	°F.	°F·
June 22 ..	69	55	− 14
23 ..	61	55	− ·6
24 ..	67	55	− 12
25 ..	70	55	− 15
26 ..	70	55	− 15
27 ..	74	55	− 19
Dec. 21 ..	41	48	+ 7
22 ..	38	47·5	+ 9·5
23 ..	32	47·5	+ 15·5
24 ..	33	47	+ 14
25 ..	34	47	+ 13
26 ..	37	47	+ 10

day is obtained inscribed on the paper scale. Once the instrument has been set against a standard thermometer it gives a direct reading of temperature changes at the point where it is placed. If erected in a survey area it would provide interesting data, and even when placed in a more permanent position the information supplied can sometimes be applied to the site of a survey.

The ecological value of temperature records lies chiefly in the determination of seasonal means, and, what is

[1] Our thanks are due to Mr. W. C. Crowther of Dulwich College for supplying the data from which this table was compiled.

often equally important, of extremes, especially low extremes, which may play an unsuspected but determining part in controlling the distribution of certain species. Mean temperatures are sufficiently well represented by the nearest observatory readings, but extremes vary so much locally that only measurements on the spot are of much value.

If a thermograph is to be bought it is best to get an instrument recording with both wet and dry elements, usually called a hygrograph (not, however, of the hair type), since this gives records of both temperature and humidity on the same instrument.

7. *Drainage and Exposure*

Data under this heading are not usually susceptible of direct measurement and can only be estimated subjectively. The direction of prevailing winds may play an important part in determining the character of the flora. Trees may have a far more varied and extensive algal and lichen flora on the side which is exposed to the prevailing wind and therefore to rain. Such points should be noted on a survey site and used to indicate the direction of the prevailing wind.

The slope of the ground and the nature of the subsoil may alter the character of the vegetation. For example, if the subsoil is rocky and impervious to water, a marsh may develop at the base of any appreciable slope, or, in extreme cases, a stream may be formed.

Slope also affects temperature. South sloping land will be warmer than that sloping north, other factors being equal, because the former is more directly exposed to the sun.

Land sloping in the direction of the prevailing wind is more exposed than that sloping in the opposite direction and will probably be colder. Cold air draining down valleys or gulleys, especially if they are backed by high hills, is frequently an important cause of low winter and spring temperatures and may considerably affect the vegetation.

8. *Biotic Factors: Animals*

The effects of insects on plants, both favourable as in pollination, or inimical, are of the greatest importance and their actions often influence ecological questions. Their study, however, really involves other branches of botanical science and lies outside our immediate concern. The chief agents which can be investigated in the field by simple means are rabbits, mice and birds, which are often selectively destructive and may, for example, completely check the regeneration of woodland trees, such as the Beech. The method of investigating their importance in a given case is by enclosing a small area within fencing which will exclude the animal concerned.

Select an area which is subject to attack by rabbits and enclose a small area in a rabbit-proof fence of wire netting. The difference will soon become apparent and striking. Photographs taken at intervals will show the changes which takes place. If rabbits are an important local factor it will be seen that, in the absence of attack, the vegetation enclosed within the fence becomes far more vigorous than that which is exposed, and that many of the protected plants produce flowers and seeds.

In making such a fence it must be remembered that rabbits will burrow down to get at luxuriant vegetation, especially in time of scarcity, and the netting should be carried down at least 6 ins. into the ground and then turned outwards for 6 ins., which discourages burrowers. Rabbits can also climb large mesh netting, so that half-inch mesh should be used. The height of the netting will depend on the type of animal concerned. Three feet is enough for rabbits and for most kinds of cattle also, if sufficiently robust.

Protection against birds involves, of course, adding a roof of netting to the enclosure. Mice require wire gauze or wire cloth to exclude them. This must be deeply sunk in the ground and the enclosure must be roofed over with the same material, as the mice will climb the wire gauze.

Experiments of this kind are called *exclosures*, because the incident factor is excluded from operating.

Attention also might be paid to the effect of dung upon meadow land. In closely cropped fields it may be noticed that certain patches of tall grass remain. Careful examination generally reveals that at some previous time dung had been dropped there. The extra nitrogenous material thus made available to the grass has produced a very lush growth which is distasteful to grazing animals, who refuse to eat it. Ecologically it may be locally important for in these areas plants mature and produce seeds while in the rest of the field the grazing prevents them from doing so.

9. *Transplant Experiments*

Tests of the effect of external factors on the behaviour of vegetation may be made by transplant experiments. From a selected plant community take up with a spade a turf, about two feet square and three or four inches thick, according to the depth of the plant roots, and transfer it to another community. Here cut out a turf of corresponding size and replant it in the position of the first turf. In this way a small sample of one plant community may be introduced into another, and vice versa. Make a detailed analysis of the vegetation of the two transplants.

After a time revisit the two areas and make a similar analysis. In this way it will be possible to discover which plants have persisted and which have died out under the changed conditions. Some plants will have spread, others become reduced. Plants from outside may have spread in, and plants from the transplant may have spread into the neighbouring community.

There are two distinct questions involved in transplant experiments: firstly, how do individual species react to environmental changes, and secondly, how do these reactions affect the plant community in which they occur. Transplants are therefore best carried out between two surveyed areas of different habitat types. In such cases many of the operative factors are known, so that the response of certain species in the transplants to changed conditions can often be referred to the operation of a definite cause. The comparative effects of certain factors may be observed at work in bringing about modification in the transplant.

A detailed quadrat study of transplants, especially if it can be extended over a number of years, will yield much interesting information which can be statistically recorded showing what changes take place before a stable condition is attained.

In this chapter we have tried to indicate various lines along which experiments may be carried out on factors influencing vegetation. It must be realized, however, that, while certain experiments are suitable for class work within a survey area, others can only be done if the site is near the laboratory and can be visited at intervals over a longer period. In the latter case extended observations might be organized by a teacher for the benefit of successive generations of students.

Finally, these remarks must be regarded as suggestions rather than as comprehensive directions, and the teacher should be able to modify and amplify them according to the conditions obtaining locally.

Chapter XII

THE ECOLOGY OF THE SEASHORE

THE problems of survey work along the seashore are on the whole similar to those elsewhere, but the methods of investigation are somewhat more difficult to apply and it is necessary to refer briefly to this kind of work.

In general it may be said that comparatively little work has been done along this line even by experienced ecologists, and the suggestions set out here are largely the result of a small number of attempts made from time to time in the course of summer excursions. In this section we are concerned, moreover, not with phanerogamic vegetation, such as may occur on a salt marsh, or on a sand dune, or even with strand plants, for these can be studied by the methods already described. In this chapter it is the algal vegetation which is under consideration.

Before setting out to analyse algal vegetation along the shore it is important to enquire in advance the state of the tides and to arrange, if possible, to begin work soon after high tide, or at least on a falling tide. Spring tides are preferable to the neaps, for more vegetation is exposed at this time. The equinoctial spring tides are ideal, but the weather conditions are apt to be a trifle cool about Easter time, and since paddling, or wading, is almost inevitable, it is not the most popular season for such work. The September spring tides may, however, be quite suitable, if the holiday period permits their utilization.

The type of work which can be done will depend largely on the local conditions. It must be remembered that algal vegetation is most prolific on rocky shores, especially in protected bays, and chiefly on the West coast of the country.

The most fruitful types of work relate firstly to the

variations in the species of algae which are found at different levels, as one passes from high tide mark downwards into the sea, and secondly to the composition of the communities in rock pools, particularly where these are comparatively large and deep.

Plane Table Survey

If possible a map should be made of the area. On a rocky shore, however, it is rarely possible to set out a grid similar to that used in field survey work. Apart from anything else, the rock surface prevents pickets and grid strings from being securely fixed, and the whole layout becomes so distorted and irregular as to render mapping by this method highly inaccurate. Furthermore it is very likely that the map will not be completed before the turn of the tide, after which anything may happen to a grid survey.

The quickest, and at the same time most accurate method is to use the *Plane Table*. This fairly simple method enables the relative positions of the essential topographical and ecological features to be plotted quite quickly on a map sheet. The outlines are then added by freehand drawing aided by pacing or tape measurements.

Select a base line which traverses the whole area to be mapped. If, for example, it is intended to study a number of rock pools, make the base line longer than the distance between the two pools furthest apart. It is generally best to make the base line along the shore. In the example shown in Fig. 47 the base line was set out on a sand bar lying some fifty yards from the shore, and only covered near high tide. A picket is inserted to mark each end of the base line. On a rocky shore the pickets are held in position with the aid of a pile of stones. The ideal pickets for such positions are a couple of camera tripods, each carrying a plumb-bob, but this is perhaps a counsel of perfection. Measure the distance between them accurately.

The plane table (Fig. 42) consists of a small drawing-board provided with a central base, which is attached to the three legs of a tripod. From the centre of the under side there hangs a plumb-bob. Student patterns of plane tables are quite cheap and it is worth while getting one

Fig. 42.—A good pattern of simple Plane Table, showing the tripod, the Alidade and the Compass, turned on its side. The Spirit Level used for levelling the table is seen on the right.

from an instrument maker, as they are generally useful. Pin a sheet of drawing paper securely to the top of the table so that it will be proof against wind. Remove the picket from one end of the base line and replace it by a short wooden peg. Arrange the tripod so that the plumb-bob is immediately over the peg, and then with the aid of a small spirit level adjust the drawing board till it is truly horizontal. This is most important. It is most easily done if the top of the tripod to which the table is attached is provided with levelling screws. These are provided on properly made plane tables.

Having got the plane table level and the plumb-bob immediately over the peg, i.e. over one end of the base line, turn the table on its base so that its length lies in the same direction as the base line. Place the box compass on the table, move it until the north pole of the needle lies over the central mark of the compass scale, and rule a line along its edge, which will fix the compass bearing of the survey plan when finished.

Now consider the size of the area in relation to the size of the paper and decide on a suitable scale. In this connection a rough test should be made by pacing, to find out the distance from the base line to the furthest point which is to be included in the map. Drive in a pin on the table to mark one end of the base line. If the base line forms one side of the area to be mapped, insert this pin towards one side of the paper, if the base line runs through the centre, place the pin on the median line of the paper.

The sighting of points in a plane table survey is carried out with an *alidade*. This is simply a ruler with two sights attached. In the high-priced tables the sights may be telescopic, but all that is essential is two points, either pins or sharp nails, driven through the ruler accurately in line with its edge, by means of which it may be aligned upon any given object in the survey area. Place the alidade on the table with its edge against the first pin. Sight the picket at the other end of the base line and rule a line along the edge of the alidade.

The actual length of the base line has been measured. All that is necessary, therefore, is to measure off an appropriate length of the line just drawn, according to the predetermined scale. This gives the base line for the map. A second pin is now inserted to mark the other end

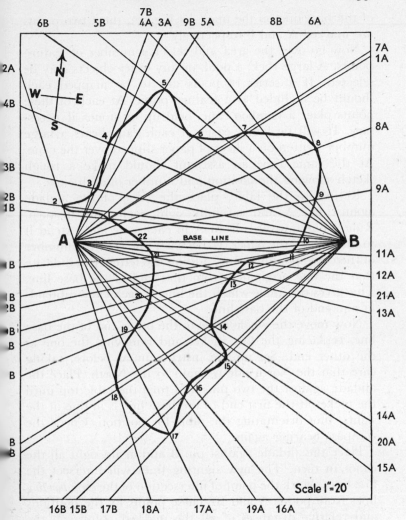

Fig. 43.—Map of a small pond at Dinas Powis, Glamorgan, made by means of a Plane Table survey. Base Line, AB, 120 ft. Twenty-two points, at the angles of the bank, were marked by canes, numbered 1–22. These were observed from each end of the Base Line. Construction lines are shown numbered at the margins.

of the base line on the map. In Fig. 43, these two points are marked A and B respectively.

Now go over the area and select a number of feature points. A large rock, a pool, or any other objects may be selected. If a series of pools are to be mapped each should be included as a feature point. At each of these points place a bamboo cane, held up by stones if necessary. It will be found easier if each cane bears a large number written on a sheet of paper slipped over the cane. At the same time an assistant should make a rough sketch showing the position and number of each cane.

Return to the table, place the edge of the alidade against pin A and carefully sight each cane in turn. Rule a line outwards from A in each case and extend it to the edge of the paper. Label each line with the number of the corresponding cane, adding the letter A to indicate that the line originated from that end of the base line. This gives the angle which the line of each cane subtends at one end of the base line.

Now move the whole table to the other end of the base line, replacing the first picket and removing the one at the other end. Set up the instrument as before, taking care that the North side is again to the North. Place the alidade against the two pins and turn the table top until the picket at the first end of the base line is in line in the sights, thus orientating the table for position. Check the compass bearing again.

Place the alidade against pin B and again sight all the canes in turn. The new sighting lines will intersect the first lines. Mark the point of intersection of the *corresponding* lines from points A and B. These points fix the positions and relative distances of all the marked points, on the scale selected, and they are used as guides in drawing the outline of the map in freehand.

Mapping with a plane table is quick and very convenient

in a difficult site like a rocky shore. It is also useful in making a general map of large areas containing a few conspicuous features. Remember, however, in the latter case, that the base line should be as long as possible.

The original map made in the field is often too dirty for final preservation. A tracing can, however, be quickly made. The position of the base line and of any points which have been specially studied should be clearly marked; the locality, date and other data marked in, and then the map should be signed by the authors.

Plane table surveys are convenient in that two, or even, if need be, one student can do the whole job, a consideration in seashore ecology where as many as possible may be needed for the algal work.

Levelling

On a seashore the environment is dominated by the tidal period, and as this depends on the height above low-water mark it is obvious that levelling has, in this case, quite a special importance. It is, however, the general slope of the shore, not the minor irregularities, which is important. For this reason it is best to keep the level at one place, preferably at one end of the base line, and move the staff about. If each feature point of the map is visited in turn, its level can be determined by direct measurement. Moreover, such a method does not involve the use of tapes, since the position of these points can be obtained from the map. The relative heights of a series of rock pools can be determined by this method.

In addition, the other end of the base line should be included in the levels measured and a perpendicular run down to low-water mark and up to high-water mark, to fix as nearly as possible the position of the base line itself between these two levels. Its actual distance from low-water mark naturally varies with every tide, but its relative position on the beach is nevertheless important.

Transects

The transition of algae from high to low tide mark is best studied by means of transects. For this purpose a modification of the simple line transect is most suitable. Since, as a rule, algae do not develop in the same profusion as land plants, it is rarely necessary to plot in all the specimens occurring along the line. In fact, it is only rarely that anything like a complete rock surface passing from high to low tide mark can be found, and where it does occur it is generally broken up by rock pools, whose vegetation, while in the main following the characteristic change of types from high to low tide, is also influenced to a considerable extent by the depth of the individual pool.

Where a suitable rock surface is found it may be possible to separate the algae into a series of distinct zones, each represented by one or more dominant species and a transect may be made to show the position and extent of these zones.

In Fig. 44 is shown one such transect. Three zones occur and the dominant and other species of algae represented are listed at the side. At the same time a diagram of the slope of the ground is shown, the results having been worked up from a survey of levels taken along the line. Such a transect gives a very good idea both of the composition of the algal vegetation and also of the transition from one type to another in relation to the length of immersion in water, which is usually the factor determining the composition of a littoral algal flora.

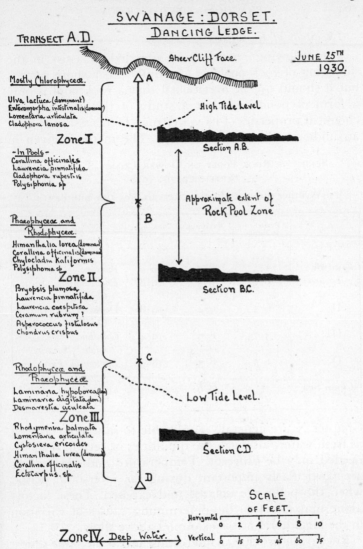

SWANAGE : DORSET.
DANCING LEDGE.

TRANSECT A.D.

Sheer Cliff Face.

JUNE 25TH
1930.

A

Mostly Chlorophyceæ.

Ulva lactuca (dominant)
Enteromorpha intestinalis (dominant)
Lomentaria articulata
Cladophora lanosa

High Tide Level

Zone I

Section A.B.

- In Pools -
Corallina officinales
Laurencia pinnatifida
Cladophora rupestris
Polysiphonia sp

Approximate extent of
Rock Pool Zone

B

Phaeophyceæ and
Rhodophyceæ.

Himanthalia lorea (dominant)
Corallina officinalis (dominant)
Chylocladia kaliformis
Polysiphonia sp

Zone II

Bryopsis plumosa
Laurencia pinnatifida
Laurencia caespitosa
Ceramium rubrum ?
Asperococcus fistulosus
Chondrus crispus

Section B.C.

C

Rhodophyceæ and
Phaeophyceæ.

Laminaria hyperborea (dom)
Laminaria digitata (dom)
Desmarestia aculeata

Low Tide Level.

Zone III

Rhodymenia palmata
Lomentaria articulata
Cystoseira ericoides
Himanthalia lorea (dominant)
Corallina officinalis
Ectocarpus sp.

Section C.D.

D

SCALE
OF FEET.

Horizontal 0 2 4 6 8 10

Zone IV Deep Water.   Vertical 0 15 30 45 60 75

FIG. 44.—Transect of Algae on the sloping surface of Dancing Ledge,
Swanage, Dorset, showing the area divided into three zones, of which
sections are shown. AD represents the line of the survey, which begins above
high tide mark and extends to below low tide level. A fourth zone of deep
water algae could not be investigated as the water depth exceeded 8 ft.

The Study of Rock Pools

The following remarks are intended to assist in the surveying of rock pools as an adjunct to a shore survey, but it should not be overlooked that a rock pool is only a form of pond and that a study of the physical and chemical properties of its water may be necessary to gain an understanding of its ecology. The methods given in

SERIES OF ROCK POOLS ON GOWER COAST, S. WALES.
SEPTEMBER 10, 1928

(Weather dull with bright intervals, low tide mid-day)

Pool	Period of isolation	Maximum pH[1]	Maximum temperature	Flora
	hours		°C	
Control	11·5	8·7	20·2	Barren.
I	10·5	9·3	19·8	*Enteromorpha*, stunted *Corallina*.
II	6·5	9·4	19·8	Dense *Ulva* and *Corallina*.
III	3·75	8·9	19·4	*Corallina, Ulva, Ceramium, Chondrus*.
IV	1·5	8·3	18·9	*Laminaria, Fucus, Corallina*, etc.
Open Sea	—	8·0	18·3	—

Chapter X are generally applicable to salt water as well as fresh water, although the quantities of certain reagents needed may be different. Temperature and pH records are particularly important during the inter-tidal period when the pools are isolated and exposed. These factors alone may be often the determining causes of variation in the algal flora, as the examples above show.

Rock pools provide an interesting and at the same time a

[1] Corrected for temperature and salt errors. The pH of the Bristol Channel is often below the normal for the open sea.

convenient method of analysing algal communities. There are several ways in which this may be done. One of the simplest is to determine by inspection the dominant types of algae present in a series of rock pools, starting at high tide mark and passing down to those which are only exposed at low spring tides. Below this level the pools cease to be distinguishable as habitats from any other rock surface under water.

Algae will often have to be collected for later determination under the microscope. The best container for small species is one of the cheap oil-skin tobacco pouches which can be rolled up and carried in the pocket. Such undetermined forms must be denoted on the transects by a symbol or a number. Little price-tickets, which can be cheaply bought in hundreds or thousands, should be tied round the collected specimens with the same symbols written on them with an ordinary soft pencil, not written in "indelible" pencil, which makes a purple mess when wetted.

A key to the seaweeds is given in Appendix II, p. 186, which should make the identification of the genera fairly easy in the field. This is a great help in making a transect list, but the specific names should also be determined later, as considerable ecological differences exist between species, even those closely related.

The pools selected should be as far as possible of uniform and moderate size and depth, and small pools a yard or two across and not more than a foot deep may be quite suitable provided they have good algal growth in them.

More accurate observations will be obtained by the use of transects (see Fig. 45). With the aid of short strings and small stones fix up a transect line along the length of the pool and make a line transect of the algae occurring every inch or two inches along the line. Since the sides

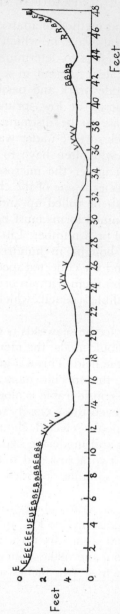

SWANAGE : DANCING LEDGE.   June 1930.

Transect of Pool.

Key. E = Enteromorpha intestinalis
      B = Bryopsis plumosa
      U = Ulva lactuca
      V = Verrucaria mucosa(?)
      R = Rhodymenia palmata

FIG. 45.—Sectional Transect of a large pool on a rock surface at Dancing Ledge, Swanage, Dorset. Depths and the contour of the bottom are shown. The position of various algal types is indicated by letters corresponding to the key.

of the pool will probably be covered with algae they should be carefully treated in the same way, as vertical surfaces. The levels across the pool can be easily obtained,

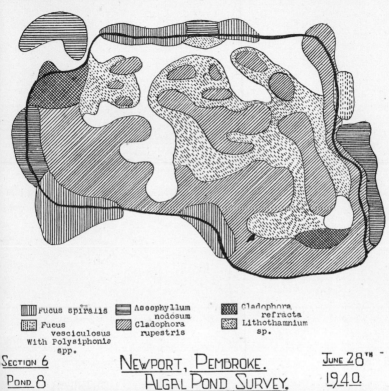

Fucus spiralis   Ascophyllum nodosum   Cladophora refracta

Fucus vesiculosus with Polysiphonia spp.   Cladophora rupestris   Lithothamnium sp.

SECTION 6   NEWPORT, PEMBROKE.   JUNE 28TH

POND 8   ALGAL POND SURVEY.   1940.

FIG. 46.—Map of the bottom of a rock pool showing the distribution of various algal types. Only the main species have been recorded. The mapping was done by the aid of a grid of strings stretched across the pool.

provided the pool is not too deep, with yard sticks thrust down to the bottom and measured to the water level. The frequency of levelling will depend on the size of the pool and its depth; in a small shallow pool every six

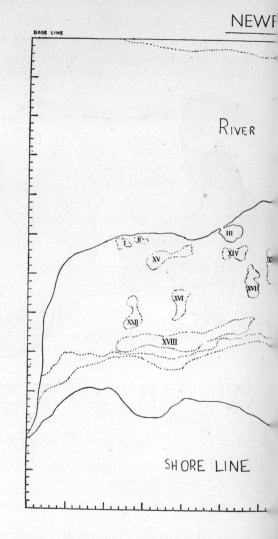

FIG. 47.—Plan of a series of Rock Pools made by mea
across the river and the positions of the pools recor
shown. The rising tide submerged the Base Line
of sequence in

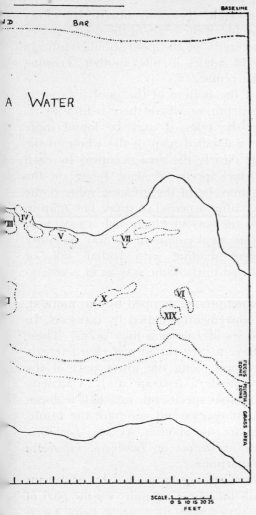

ROCK POOL   June 28th 1940

BASE LINE

ND   BAR

A WATER

III
IV
V
VII

I
X
VI
XIX

FUCUS ZONE
PLUVIAL ZONE
GRASS AREA

SCALE. 0 5 10 15 20 25
FEET

se Line set upon a sand bar. Grid strings were taken
osition of certain algal zones on the foreshore is also
pools before the survey was completed, hence the lack
bering of the pools.

inches should be enough. Try and obtain by closer measurements the profile of the edges of the pool.

A really complete study of the pool should embrace two transects at right angles to one another, crossing approximately at the centre.

Where the area of the bottom of the pool is large in comparison with its depth, or where there is little or no algal vegetation on the sides, it may be found more convenient to make a detailed map of the whole of the pool and to indicate clearly the area occupied by each species, or by the chief species of algae living on the bottom. Such maps may be of the greatest value if the same pool is revisited after a period of time. The map in Fig. 46 shows clearly the way in which such a map may be made. The separate algal communities may be either indicated by different shading with Indian ink, or colours can be employed in the same way as in a vegetation map.

Deep pools can sometimes be mapped by this method, but they are more conveniently studied by transects. In such pools large species of *Laminaria* may occur. These large wracks often have a very varied and characteristic Red Algal flora growing along the lower part of the stipe, which may be made a separate study. Calculate the percentage of different species on, say, half a dozen stipes of the same host species and compare the results with a similar estimate on another species. *Laminaria digitata* and *Laminaria saccharina* or *Laminaria hyperborea* make interesting comparisons.

The problem of working a deep pool is by no means easy to solve and calls for some ingenuity on the part of the investigators. A water-glass[1] is a great help, for it enables the vegetation to be viewed without trouble from surface reflection. Another helpful device is a small hand

[1] See Chapter X, p. 114.

mirror, which can be very useful in illuminating dark corners. Levelling may be possible with the aid of a picket, the inches being measured with a yard-stick tied to the appropriate part of the picket, which is then thrust into the water. In water over 5 ft. deep, sounding with a weighted measuring string is the only accurate way. It will be realized, however, that in water over 4 ft deep any accurate transect work is impossible, and in deeper water, even if any algae are present, they can only be brought up at random with a grapnel or by diving for them. Such random sampling, therefore, becomes little more than collecting, and is outside the province of this book.

When rock pools are surveyed in relation to one another as a series down a shore (see Fig. 47), it is desirable to relate the levels to one another. This can be best done by fixing the optical level in the highest pool and determining its actual height above the water in the pool by direct measurement. Now pass from pool to pool with the staff, placing it on the bottom at the intersection of the two transects. Note the height of the water surface in the pool and read off the height visible in the level. In this way a relation between the water levels in all the pools can be found. In this work it is best, if possible, for the level itself to be kept at one point all the time, and only the staff moved about.

Another method may be used if no level is available, but it involves more trouble. The time at low tide must be accurately observed. The rise of the tide is then followed and the time at which it reaches each pool is noted. This gives the relative height of each pool, in relation to the others and to the total rise of the tide in the time between high and low tide, i.e. 6 hrs. 18 mins. The calculation is not, however, a simple proportion, as the tidal movement is not constant but harmonic, that is to say, it accelerates up to half tide and decelerates or retards from then onwards to high tide. The simplest solution is by means of a graph (see Fig. 48). The two parallel lines AX and BY represent high water and low water levels respectively, the time interval between which is 376 mins. Suppose the tide leaves a certain pool 1 hr. 59 mins. after high tide. This is $\dfrac{119}{376}$ of the whole

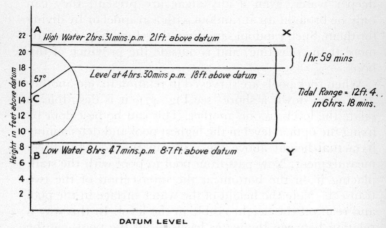

DATUM LEVEL

FIG. 48.—Diagram showing the method of calculating the relative heights of rock pools by observing the time at which the rising tide enters them. For details see in text.

period. This fraction of 180° is 57°. From the centre of the circle ABC draw a line at the angle of 57°. From the point at which this cuts the circle draw a parallel to AX. This gives the height of the pool between high and low water level.

If the total tide rise can be measured or ascertained from the published tables, this enables one to find exactly the height of each pool above low tide mark, but even without this figure it fixes the *relative* positions of the pools on the shore, which is often sufficient as a basis of comparison between them.

Rock Surfaces

It is sometimes possible to make an interesting analysis of algae growing either on vertical or horizontal rock surfaces. Where vertical cliffs of fairly solid rock rise straight out of the water it may be possible to prepare a vertical transect. Starting as deep in the water as possible the successive types of algae are recorded in relation to their depth below the surface at low tide mark.[1] The types of algae left bare are then treated in the same way, and finally, by direct observation, if possible, the position of high tide mark is also recorded. From tide tables it is easy to discover the height of the tide on any particular day and from this the position of the equinoctial spring tides can be estimated.

Attention should also be given to the various characteristic lichen zones which are often very clearly shewn on such rock surfaces near and above high tide mark, that is to say, in the "spray" zone. Just above the *Pelvetia* zone is the characteristic black *Verrucaria* zone, consisting of *V. maura* and *V. marina*. Higher up, above high tide mark, there is another black zone of the minute shrubby form, *Lichina pygmaea*, especially in the West and North. Higher still comes the conspicuous orange-yellow zone of *Xanthoria pariertina*, and lastly the grey tufts of *Ramalina scopulorum*, which may reach even to the top of the cliffs.

Horizontal rock surfaces are less common, but where one is found it should be examined in detail or a series at different levels may be compared and the relation of the algal flora to period of exposure studied (see Fig. 49). It

[1] Loose diving helmets, provided with an air-supply from a tyre-inflator, have been constructed by some amateurs of under-water surveying, especially in the United States. They speak enthusiastically of their use.

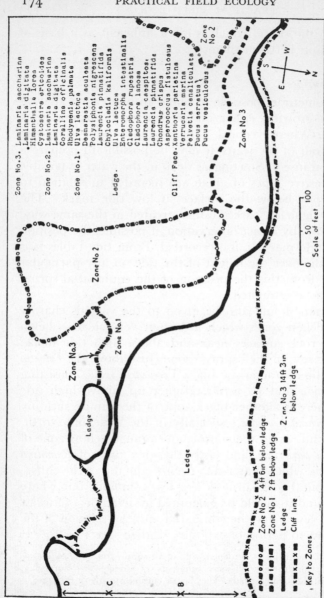

FIG. 49.—Map of Rock Surfaces, Dancing Ledge, Swanage, June, 1930. The shore consists of a series of ledges lying at different heights from one another and carrying distinct algal communities. High tide mark is shown by the cliff line, while zone No. 3 is below low tide mark and therefore always covered by the sea. There is a general slope southwards, as will be seen from Fig. 44, which shows the line transect taken along the line AD. The chief algae occurring in each zone are tabulated on the right of the map.

will be realized, however, that such flat rock surfaces, of suitable consistency, occur only rarely, where the successive bedding planes of the rock lie roughly horizontal, or on limestone coasts where erosion by solution levels down the rocks of the foreshore. Igneous or metamorphic rocks scarcely ever provide surfaces flat enough for quadrat surveys.

Zonation parallel to the coast line is a phenomenon which can be observed among the algae of almost any shore. Comparatively little is known of the facts except in a few cases, and there is plenty of room for original observations along these lines, which may have a distinct scientific value, if care is taken to see that the species are correctly determined.

There is likewise much to be learned about the associations of algae in the pools and their dependence on environmental factors. If interesting observations have been made it is well worth while taking a little trouble to get the identifications of the species verified. The staff of the Botany Department in the nearest University will probably be able to help in this, or, if not, they will be able to tell an enquirer where to apply.

When submitting specimens for identification, do not send a wet, semi-decayed mush of evil-smelling wrack, but neat dried specimens mounted on stiff paper. Float the specimens in *fresh* water in a shallow dish, introduce a sheet of paper below them as they float, arrange them carefully with brush and needle to display their full beauty, and then carefully draw out paper and specimen together. Cover the specimen with a piece of muslin and press it *lightly* in a drying press between blotting papers until it is quite dry. In this way even the most delicate species can be well preserved and displayed.

From what has been said, it will be seen that a rocky seashore offers many opportunities for ecological survey work. That the work is by no means as easy as land surveying will readily be appreciated. Moreover, the nature of the strata varies so much in different localities, producing such widely different conditions, that it is impossible to give directions as precisely as in land surveys. Each locality must be studied separately in the way most suited to the local conditions. The foregoing descriptions, however, may help to suggest methods of approach to the problem, which a good student can modify for the particular conditions obtaining in each case. Comparatively little work of this kind has been done, and simple quantitative and statistical data collected in this way may furnish important contributions to Science.

Chapter XIII

THE ECOLOGICAL HERBARIUM

ALTHOUGH the compilation of a general herbarium scarcely comes within the province of field ecology, there is one aspect of such work which is directly related to it. It is obviously impossible, in the time which a student can devote to field work, for him to study, or even see, more than a limited number of plant communities. It may be possible, however, over a number of years, by careful selection of suitable localities, for the teacher to visit a much more extensive range of communities. If representative plants are collected from each of these as they are visited, it ought to be possible to build up an herbarium of considerable value in field studies.

The object of this herbarium should be ecological rather than systematic and therefore it is only the commoner plants which are of importance. It should be easy in the course of any survey work to determine, from the valance analysis and from the belt transect histogram, which are the dominant and sub-dominant species of the community, and it is specimens of these plants which should form the foundation of the collection.

The plants should be dried in the usual way and then mounted on herbarium sheets of standard size (16 ins. by 11 ins). each of which must then be labelled. The labels used in such work differ somewhat from those used in a systematic herbarium as will be seen from the example given in Fig. 50.

It will be seen that emphasis is laid on the ecological rather than on the systematic position of the plant, for even the Family is omitted while the ecological position of the plant is stressed in detail. For the same reason the sheets are not arranged as in a systematic herbarium but

M

according to the plant communities. The degree of classification must necessarily depend upon the total size of the herbarium. At first it may be sufficient to use a separate folder to enclose only the larger ecological divisions; thus using only a single folder for all the sheets

---

UNIVERSITY COLLEGE: CARDIFF

Department of Botany

ECOLOGICAL HERBARIUM

Name ....

Association ....

Society ....

Locality ....

Date..Abundance.............per cent

Collector ....

Fig. 50.—An Ecological Herbarium Sheet Label, showing the headings which, when filled in, make the drafting of the sheet into the herbarium a simple matter, as well as thereby supplying valuable data.

of salt-marsh plants, another for sand dune plants, and so on. As the collection grows, further subdivision may become desirable; thus the *Salicornia* association may be separated from the *Glyceria* association into separate folders. The use of different coloured folders may be a considerable help in this arrangement. If say, brown folders are used to embrace all the salt-marsh plants, green folders for the separate associations, and white ones for the individual societies, an elaborate but very instructive herbarium will in time accumulate.

A study of such an herbarium will be of great value. It can be used as a teaching collection to familiarize students with the characteristics of communities which they are unable to visit, and it can also be employed for instruction, before a locality is visited, on the types of plants likely to be met with and those which will probably be found to be ecologically most important.

It will be realized that several sheets of the same plant will have to be included in different parts of the collection, for many plants are characteristic of more than one plant community. On the other hand, since the herbarium should comprise chiefly common plants, a comparatively complete collection can be amassed within a reasonable time. Moreover, since such a collection is concerned with dominant species rather than with rarities, it is well within the compass of student work. For the same reason critical determination of the species in difficult genera is less often required than it is in a systematic herbarium.

While it is important that representative plants showing flowers and fruits should be included, it is equally important that specimens in the juvenile condition should be pressed. Frequently in ecological work it is necessary to name a plant, not from a fully grown specimen, but from a seedling. Reference to a good ecological herbarium may then be the only means of identification short of growing the seedling to maturity.

In a herbarium of this kind not only flowering plants, but also mosses, liverworts, ferns and fungi should be represented. Mosses particularly will occupy an important place, especially in certain sections. If the herbarium is extended to cover maritime work, collections of seaweeds must be included, though considerable original research will probably be required before any satisfactory arrangement of seaweeds into ecological associations can be reached.

Few really good ecological herbaria exist at the present time, even in the Universities, for the influence of systematists is still strong and considerations of space alone often preclude the compilation of two separate herbaria. On the other hand a good working ecological herbarium need not occupy nearly as much room as a systematic one, for, as has been pointed out already, only the more characteristic species need be considered, at all events in the first instance. Enterprising schools might very well compile a collection at least from their own immediate district, extending it further as opportunity arises.

Finally, as a means of fixing in the mind of a student the appearance of the plants which form the more important components of a community he has been investigating, the collecting, drying and mounting of specimens, is of the utmost value, quite apart from the satisfaction derived from making a contribution to the permanent records in his school or college.

It is not proposed here to describe in detail the methods of drying and mounting such herbarium specimens, for the methods are identical with those commonly used in systematic work. A detailed description of the process employed for Phanerogams and also the special precautions necessary in dealing with Cryptogams are fully dealt with in our *Text Book of Practical Botany*.[1]

[1] In the Press.

Appendix I

KEY FOR DETERMINING SPECIES OF BRITISH GRASSES BY VEGETATIVE CHARACTERS

1. Leaf needle-like, not easily flattened out,   2.

   Leaf flat, or if needle-like, easily flattened out,   7.

2. Leaf hair-like, edges inrolled,   *Deschampsia flexuosa.*

   Leaf distinctly folded. Section V-shaped,   3.

3. Plants creeping, perennial,   *Glyceria maritima.*

   Plants densely tufted, perennial,   4.

   Plants not densely tufted, annual,   6.

4. Leaves spreading at right angles to whitish sheaths, ligule short,   *Nardus stricta.*

   Leaves not at right angles to sheath, ligule obsolete or wanting,   5.

5. Lower leaf sheaths reddish,   *Festuca rubra.*

   Lower leaf sheaths not reddish, split half way to base,   *Festuca ovina.*

6. Leaf sheaths rough,   *Aira caryophyllea.*

   Leaf sheaths glabrous,   *Aira praecox.*

7. Leaf folded in bud, stem more or less flat,   8.

   Leaf rolled in bud, stem round or square,   22.

8. Small auricles at base of leaf-blade, basal sheath glossy, ligule very short,   *Lolium perenne.*

   No auricles at base of leaf-blade,   9.

9. Plants aquatic, leaves large,   10.

   Plants not aquatic,   11.

10. Leaf sheath keeled, blade prominently ribbed,   *Glyceria fluitans.*

    Leaf sheath not keeled, blade not prominently ribbed,   *Glyceria aquatica.*

11. Leaf with whitish median lines, hooded
     at tip,   12.

    Leaf without whitish median lines or
     hooded at tip,   15.

12. Leaf stiff, glaucous, ligule very short,   *Sesleria coerulea.*
    Leaf not stiff or glaucous,   13.

13. Stem rough, ligule long,   14.
    Stem smooth, ligule short,   *Poa pratensis.*

14. Blade shining below,   *Poa trivialis.*
    Blade dull below,   *Poa annua.*

15. Sheath or blade or both hairy,   16.
    Sheath and blade smooth,   17.

16. Ligule membranous, long,   *Avena pubescens.*
    Ligule a tuft of hairs,   *Triodia decumbens.*
    Ligule short or obsolete,   *Bromus erectus.*

17. Ligule long,   18.
    Ligule short,   21.

18. Ligule bifid, leaf usually rolled: sand
     dunes,   *Ammophila arenaria.*

    Ligule not bifid, leaf usually flat,   19.

19. Ligule embracing the stem,   *Briza media.*
    Ligule not embracing the stem,   20.

20. Leaf with very prominent ribs, ligule
     pointed,   *Deschampsia caespitosa.*

    Leaf ribs not prominent, ligule ragged,
     shoot flattened,   *Dactylis glomerata.*

21. Ligule blunt, leaf sheaths sharply keeled,   *Poa compressa.*

    Ligule blunt, notched, basal sheaths
     yellowish,   *Cynosurus cristatus.*

    Ligule acute,   *Avena pratensis.*

22. Pointed projections opposite base of leaf-
    blade, stem square,   *Melica uniflora.*

    No pointed projections at base of leaf-
    blade, stem not square,   23.

23. Auricles at base of leaf-blade,   24.
    No auricles at base of leaf-blade,   31.

24. Lowest leaf-sheath pink, glossy,   25.
    Lowest leaf-sheath not pink or glossy,   26.

25. Margin of leaf at base smooth, veins
    indistinct,   *Lolium italicum.*

    Margin of leaf at base rough, veins
    distinct,   *Festuca elatior.*

    Leaf all rough. Plants very tall; woods *Bromus giganteus.*

26. Basal sheath hairy,   27.
    Basal sheath smooth or nearly so,   28.

27. Hairs long, ligule prominent; woods,   *Bromus asper.*
    Hairs short, ligule very short; pastures, *Hordeum pratense.*

28. Ligule with hairs at base, auricles very
    short,   *Anthoxanthum odoratum.*

    Ligule hairless,   29.

29. Ligule short,   30.
    Ligule a mere rim. Markedly stoloni-
    ferous,   *Agropyron repens.*

30. Leaf thick, ridged, glaucous,   *Elymus arenarius.*
    Leaf thin, papery, downy,   *Hordeum murinum.*
    Leaf short, stiff, rough,   *Lepturus filiformis*

31. Lower leaf sheaths hairy,   32.
    Lower leaf sheaths smooth or nearly so, 39.

32. Basal sheaths white, with pink veins,   33.
    Basal sheaths not white, without pink
    veins,   34.

33. Leaf sheaths inflated,   *Holcus lanatus.*
     Leaf sheaths not inflated,   *Holcus mollis.*

34. Ligule prominent, membraneous,   35.
     Ligule inconspicuous,   38.

35. Leaf sheath keeled,   36.
     Leaf sheath not keeled,   *Trisetum flavescens.*

36. Leaves thin and papery, sparsely covered
     with long hairs, yellowish-green,   *Brachypodium sylvaticum.*
     Leaves rough, rigid and erect, sparsely
     hairy, tending to roll up,   *Brachypodium pinnatum.*
     Leaves downy,   37.

37. Upper sheaths smooth,   *Bromus sterilis.*
     Upper sheaths downy,   *Bromus mollis.*

38. Leaf glaucous, purple at the tip,   *Molinia coerulea.*
     Leaf sheath and blade downy,   *Koeleria cristata.*

39. Plants aquatic,   40.
     Plants not aquatic,   41.

40. Ligule a tuft of hairs,   *Phragmites communis.*
     Ligule membraneous,   *Phalaris arundinacea.*

41. Woodland grass, ligule large,   *Milium effusum.*
     Pasture grass,   42.

42. Lowest leaf sheaths coloured,   43.
     Lowest leaf-sheaths not coloured, 44.

43. Lowest leaf sheaths brown-violet or
     black,   *Alopecurus pratensis.*
     Lowest leaf sheaths yellow,   *Cynosurus cristatus.*

44. Leaf prominently ribbed,   45.
     Leaf not prominently ribbed,   49.

45. Perennial,   46.

    Annual, usually a weed of cultivated
    land,   *Alopecurus agrestis.*

46. Stem constantly bent at nodes, blade
    sharply ribbed,   *Alopecurus geniculatus.*

    Stem not constantly bent, leaves taper-
    ing to base,   47.

47. Ligule oblong,   48.

    Ligule not oblong,   49.

48. Leaves flat, smooth,   *Agrostis canina.*

    Leaves needle-like, margins involute,
    rough,   *Agrostis setacea.*

49. Ligule short, truncate,   *Agrostis tenuis.*

    Ligule long, acute,   *Agrostis alba.*

    Ligule long, acute, plant vigorously
    creeping,   *Agrostis stolonifera.*

50. Leaf-stalk prominently keeled, leaf dark
    green,   *Arrhenatherum elatius.*

    Leaf-stalk scarcely keeled except at top,
    leaf light green,   *Phleum pratense.*

Appendix II

KEY TO THE CHIEF GENERA OF BRITISH MARINE ALGAE

(By the late PROFESSOR R. J. HARVEY GIBSON)

A. CHLOROPHYCEAE

1. Fronds unicellular, epiphytic,   *Codiolum.*
   Fronds wholly ceonocytic,   2.
   Fronds incompletely coenocytic,   4.
   Fronds multicellular,   7.

2. Brackish water,   *Vaucheria.*
   Marine,   3.

3. Filaments free, branched,   *Bryopsis.*
   Filaments interwoven into a spongy mass,   *Codium.*

4. Fronds filamentous, unbranched,   5.
   Fronds filamentous, branched,   6.

5. Fronds delicate, flaccid,   *Urospora.*
   Fronds coarse, more or less rigid,   *Chaetomorpha.*

6. Branches few, with colourless hooks,   *Rhizoclonium.*
   Branches very numerous,   *Cladophora.*

7. Fronds endo- or epiphytic,   8.
   Fronds free,   10.

8. Fronds discoid, epiphytic,   *Pringsheimia.*
   Fronds filamentous, branched,   9.

9. Endophytic in the cell walls of Algae, Zoo-
   phyta and Bryozoa,   *Endoderma.*
   Endophytic in the shells of Mollusca,   *Gomontia.*

10. Fronds filamentous, unbranched,   *Ulothrix.*

    Fronds unilamellar when adult,   11.

    Fronds bi-lamellar or tubular when adult,   12.

11. Fronds unilamellar at all times,   *Prasiola.*

    Fronds at first tubular or globular,   *Monostroma.*

12. Fronds tubular (or nearly so) at all times,   *Enteromorpha.*

    Fronds bi-lamellar,   *Ulva.*

B. Phaeophyceae

1. Fronds gelatinous, irregularly globose, hollow,
   composed of closely packed chains of
   minute spherical cells,   *Leathesia.*

   Fronds coriaceous, irregularly globose,
   hollow, walls composed of large vesicles,
   covered with minute cells in parallel rows,   *Colpomenia.*

   Fronds discoid or saucer-shaped, stipitate,
   with long strap-like receptacles,   *Himanthalia.*

   Fronds in the form of adherent crusts or
   pellicles,   2.

   Fronds consisting of a compact basal region   { *Elachistea.*
   with erect, free, monosiphonous filaments,   { *Myriactis.*

   Fronds elongate tubular,   4.

   Fronds elongate, solid or becoming partly
   tubular with age,   6.

   Fronds elongate, filamentous, branched,
   sometimes with basal rhizoidal filaments,   15.

   Fronds erect, membranous or ribbon-shaped,
   more or less expanded,   17.

2. Fronds thick, compact, parenchymatous,   *Ralfsia.*

   Fronds minute, thin, formed of an encrusting
   layer of cells and short vertical monosiphonous filaments, with or without
   colourless hairs,   3.

3. Colourless hairs present,   *Ascocyclus.*

   Colourless hairs absent,   *Myrionema.*

4. Fronds delicate, unbranched, tubular throughout,   5.

   Fronds delicate, branched, tubular, sori in whorls at short intervals,   *Striaria.*

5. Fronds tubular but constricted at intervals, sporangia generally diffused,   *Scytosiphon.*

   Fronds tough, cylindrical, tubular, but with diaphragms at intervals,   *Chorda.*

   Fronds irregularly swollen, sporangia in sori,   *Asperococcus.*

6. Fronds slimy, with a central strand of hyphae and a cortex of short horizontal filaments,   7.

   Fronds composed of prosenchymatous medullary cells, becoming smaller towards the surface, with or without an axial hypha,   9.

   Fronds composed of cells of nearly uniform length, arranged in transverse rows,   13.

7. Fronds tough and dense,   *Chordaria.*

   Fronds soft and flaccid,   8.

8. Gametangia immersed in the cortex,   *Castagnea.*

   Gametangia external to the cortex,   *Mesogloia.*

9. With axial monosiphonous filament,   10.

   Without axial monosiphonous filament,   11.

10. Sporangia imbedded, frond tough,   *Desmarestia.*

   Sporangia in stalked, clavate receptacles, frond soft to cartilaginous,   *Sporochnus.*

11. Fronds unbranched, covered with hyaline hairs,   *Litosiphon.*

   Fronds branched,   12.

12. Sporangia in projecting sori,   *Stilophora.*

  Sporangia scattered and immersed, fronds beset with hyaline multicellular hairs,   *Dictyosiphon.*

  Sporangia scattered and immersed, fronds without hyaline hairs,   *Stictyosiphon.*

13. Fronds minute, monosiphonous below, beset with short branches and hyaline hairs,   *Myriotrichia.*

  Fronds polysiphonous throughout, but ending in prominent apical cells,   14.

14. Fronds setaceous, branching usually alternate,   *Sphacelaria.*

  Branching regularly pinnate and opposite,   *Chaetopteris.*

  Branches densely covered with short ramuli, more or less regularly whorled,   *Cladostephus.*

  Fronds with shaggy bases of descending filaments, pinnate, alternate, axes with four rows of elongate cells,   *Halopteris.*

  Fronds with shaggy bases of descending filaments and erect bushy branches,   *Stypocaulon.*

15. Sporangia and gametangia stalked, free,   *Ectocarpus.*

  Sporangia and gametangia intercalary in the branches,   *Pylaiella.*

  Sporangia immersed, branches opposite,   16.

16. Sporangia solitary or in pairs, replacing lateral ramuli,   *Isthmoplea.*

  Sporangia imbedded in branches at the points of origin of two opposite lateral ramuli,   *Phaeospora.*

17. Fronds stipitate, lamina unbranched, rarely proliferous,   18.

  Fronds stipitate, lamina branched,   21.

18. Midrib present,   *Alaria.*

  Midrib absent,   19.

19. Fronds thin, subsessile,　　　　　　　　　　20.

Fronds thick, somewhat leathery, puckered, stipitate,　　　　　　　　　*Laminaria* (in part)

20. Sporangia, scattered over the whole frond,　　*Phyllitis.*

Sporangia in sori,　　　　　　　　　　　　*Punctaria.*

21. Fronds large leathery, midrib absent, sporangia in broad patches over the lamina,　　22.

Fronds somewhat delicate and membranous, sporangia in sori, midribs present or not,　　23.

Fronds large, variable in size, leathery, sporangia absent, gametangia in conceptacles,　　　　　　　　　　　　　　　25.

22. Stipe compressed, twisted, with warty and hollow base,　　　　　　　　　　*Saccorhiza.*
Stipe cylindrical with branched hapters,　　*Laminaria* (in part)

23. Gametangia and sporangia sessile,　　　　24.
Fronds multifid, sporangia and gametangia stalked,　　　　　　　　　　　　　*Cutleria.*

24. Fronds dichotomously branched, without midribs, sporangia or gametangia sessile in sori,　　　　　　　　　　　　　*Dictyota.*

Fronds dichotomously branched, with midribs, sporangia in sori,　　　　　*Dictyopteris.*

Fronds fan-shaped, aciniate, sporangia in concentric bands,　　　　　　　　*Taonia.*

Fronds fan-shaped, entire or lobed, margins ciliate, sporangia in concentric bands,　*Padina.*

25. Conceptacles in terminal receptacles,　　　26.

Conceptacles in lateral receptacles, air-floats simple intercalary in the frond,　　*Ascophyllum.*

Conceptacles in lateral receptacles, air-floats lateral stalked, chambered,　　　*Halidrys.*

26. Midrib present,　　　　　　　　　　　*Fucus.*
Midrib absent,　　　　　　　　　　　　27.

27. Fronds cylindrical,   28.

    Fronds canaliculate,   *Pelvetia.*

28. Fronds smooth without air-floats,   *Bifurcaria.*

    Fronds covered with short spine-like ramuli,
    air-floats beneath the receptacles,   *Cystoseira.*

C. RHODOPHYCEAE

1. Fronds calcareous,   2.

    Fronds non-calcareous,   4.

2. Fronds erect, branching, jointed,   *Corallina.*

    Fronds thin, encrusting,   3.

    Fronds thick, massive encrusting, rising at   { *Lithothamnion.*
    intervals into irregular prominences,   { *Phymatolithon.*

    Fronds minute nodular, endophytic on
    Corallina,   *Choreonema.*

3. Fronds epiphytic, delicate, smooth,   *Melobesia.*

    Fronds epiphytic, pustulate,   *Dermatolithon.*

    Fronds partially adherent to rock, foliaceous, *Lithophyllum.*

4. Fronds filamentous, endophytic in cell-walls
    of *Cladophora pellucida*,   *Schmitziella.*

    Fronds encrusting,   5.

    Fronds erect, free,   6.

5. Fronds cartilaginous, tetrasporangia and
    cystocarps in conceptacles, tetraspores
    zonate,   *Hildenbrandtia.*

    Fronds cartilaginous, tetrasporangia in warts
    between vertical filaments, tetraspores
    cruciate,   *Cruoriella.*

    Fronds gelatinous, tetrasporangia inter-
    calary in vertical filaments, tetraspores
    cruciate,   *Petrocelis.*

    Fronds gelatinous, tetrasporangia arising
    laterally from vertical filaments, tetraspores
    zonate,   *Cruoria.*

6. Fronds filamentous, monosiphonous at least
   in the younger branches,   7.

   Fronds cylindrical, solid or tubular,   17.

   Fronds more or less compressed or expanded,
   membranous,   27.

7. Fronds unbranched,   8.

   Fronds branched,   9.

8. Fronds monosiphonous throughout, crimson,   *Erythrotrichia.*

   Fronds monosiphonous at base, polysiphonous
   upwards, purple,   *Bangia.*

9. Fronds monosiphonous throughout,   10.

   Fronds arising from a disc, monosiphonous in
   younger branches, corticated by filaments
   in main axes,   16.

   Fronds arising from a disc, composed of single
   axial chains of large elongated cells, corticated by minute cells at each node, apices
   incurled,   *Ceramium.*

10. Fronds minute, usually epiphytic,   11.

    Fronds fairly large,   13.

11. Fronds erect, rising from creeping filaments,
    tetrasporangia with four tetraspores,   12.

    Fronds erect arising from a disc, tetraspores
    solitary,   *Acrochaetium.*

12. Fronds irregularly branched, tetraspores
    cruciate,   *Rhodochorton.*

    Fronds alternately branched, pinnate, tetra-
    spores tetrahedral,   *Spermothamnion.*

13. Fronds arising from a disc,   14.

    Fronds arising from a fibrous base,   15.

14. Fronds oppositely branched, tetraspores cruciate, *Antithamnion.*

Fronds irregularly branched, tetraspores tetrahedral, { *Griffithsia.* *Monospora.*

Fronds with whorled branches, tetraspores tetrahedral, *Halurus.*

Fronds with regularly alternate, pinnate branches, tetraspores tetrahedral, *Compsothamnion.*

15. Fronds with whorled branches, *Sphondylothamnion.*

Fronds with mostly dichotomous branches, *Bornetia.*

Fronds pinnately branched, tetrasporangia with numerous tetraspores, *Pleonosporium.*

16. Fronds densely and irregularly branched, corticated with fine filaments in lower axes, tetrasporangia solitary, *Callithamnion.*

Fronds pinnately branched, corticated with fine filaments in lower axes, tetrasporangia in serial rows, *Seirospora.*

17. Fronds solid, 20.

Fronds tubular, 18.

18. Tube segmented by diaphragms, regularly constricted without monosiphonous ramuli, { *Lomentaria.* *Chylocladia.* *Champia.*

Fronds not constricted, with monosiphonous lateral ramuli, *Spyridia.*

Tube continuous throughout, 19.

19. Wall composed of elongated parenchymatous cells becoming smaller outwards. Brown-purple, *Dumontia.*

Wall composed of interwoven hyphae with peripheral dichotomously branched moniliform filaments. Bright crimson, *Gloiosiphonia.*

20. Root fibrous, creeping, 21.

Root discoid, compact, 22.

N

21. Fronds composed of core of loose reticulate hyphae, cortex of dichotomous chains of cells. Fronds constricted at intervals, *Catenella.*

Fronds composed of closely packed core of hyphae, cortex parenchymatous. Fronds becoming smaller-celled upwards, *Cystoclonium.*

Fronds composed of short core of hyphae, becoming shorter outwards, apices of branches circinately inrolled, *Bostrychia.*

Fronds dichotomously branched, composed of central core of hyphae, inner cortex parenchymatous, outer cortex of horizontal moniliform threads, *Furcellaria.*

Fronds composed of central parenchyma, cells becoming smaller outwards, { *Gracelaria.* *Chondria.* }

22. Fronds composed of medullary hyphae, cortex of horizontal dichotomous moniliform filaments, 23.

Fronds composed of medullary prosenchyma, cortex of horizontal dichotomous moniliform filaments, 24.

Fronds dichotomously branched, composed of medullary hyphae, inner cortex of parenchyma, outer cortex of horizontal moniliform filaments, *Polyides.*

Fronds composed of large celled medullary parenchyma and small celled cortex, 25.

Fronds composed of parallel chains of cells of uniform length in each segment, 26.

23. Fronds dichotomously branched, branches of uniform thickness, blunt apices. Purple-brown, *Nemalion.*

Fronds of one chief axis, tapering from centre to base, branches alternate, pointed. Dull pink, *Helminthocladia.*

24. Fronds gelatinous, main axis and branches of irregular thickness. Pale red, *Helminthora.*

Fronds cartilaginous, dichotomously branched.
Purple-black, *Gymnogongeus.*

Fronds hard and wiry, irregularly branched.
Purple-black, *Ahnfeltia.*

25. Fronds gelatinous, main axis irregularly
branched, beset with minute spindle-
shaped ramuli, *Naccaria.*

Fronds soft and sub-cartilaginous, alternately
pinnate, branches with obtuse ends, *Laurencia.*

Fronds cartilaginous, main axis alternately
branched, terminal ramuli opposite, pinnate inrolled, *Halopitys.*

Fronds soft, irregularly branched, closely set
with monosiphonous, dichotomous, pink
ramuli, *Dasya.*

Fronds soft, compound, pinnately branched,
ramuli deciduous leaving rough stumps.
Brownish-red, *Rhodomela.*

26. Fronds polysiphonous, occasionally with { *Polysiphonia.*
apical branched monosiphonous ramuli, { *Pterosiphonia.*

Fronds polysiphonous, beset throughout with
branched monosiphonous ramuli, *Brongniartella.*

27. Fronds membranous, expanded, entire, sessile
or with very short stipe, 28.

Fronds membranous, expanded lobed or
divided, 29

Fronds compressed, narrow almost filamentous, 32.

Fronds stalked, leafy, 35.

Fronds narrow, ribbon-like of uniform
breadth, 36

Fronds lobed, membranous, stalked, cartilaginous, 37.

28. Root discoid, fronds sessile, entire, delicate
    membranous, 1–2 cells thick,   *Porphyra.*

    Root discoid, fronds cartilaginous, mem-
    branous, obovate, composed of central
    hyphae and moniliform cortical filaments,   *Dilsea.*

    Root discoid, fronds fleshy membranous,
    obovate composed of parenchymatous
    central hyphae and cortex becoming small-
    celled outwards,   *Schizymenia.*

29. Root discoid,   30.

    Root fibrous,   31.

30. Fronds gelatinous membranous, irregularly
    divided, central loose network of hyphae
    with superficial layer of minute cells,   *Halarachion.*

    Fronds soft membranous, stipitate, palmate,
    ribbon-like segments, central parenchyma
    becoming smaller outwards,   *Stenogramme.*

    Fronds thick membranous, branching dicho-
    tomous, substipitate, palmate, parenchy-
    matous with anastomosing lacunae,   *Callophyllis.*

    Fronds delicate, membranous, lobed or bifid,
    often ciliate, composed of uniform paren-
    chyma,   *Nitophyllum.*

31. Fronds delicate membranous, branching
    dichotomous, veinless, parenchymatous,
    tetraspores zonate,   *Rhodophyllis.*

    Fronds subcartilaginous, irregularly lobed,
    ciliate margins, large central and small
    cortical and peripheral cells,   *Calliblepharis.*

32. Root discoid,   34.

    Root fibrous,   33.

33. Fronds cartilaginous, compressed, irregularly
    branched with small leafy expansions,
    central hyphae and parenchymatous cortex, *Gelidium.*

Fronds compressed, alternate narrow branches, ultimate ramuli pectinate, central parenchyma becoming smaller towards the surface, — *Plocamium.*

34. Fronds soft, compressed, narrow terminal branches with opposite subulate ramuli, replaced on alternate sides by cystocarps, — *Bonnemaisonia.*

Fronds cartilaginous, compressed, narrow branches fringed with cilia with cystocarps, central hyphae, parenchymatous cortex, superficial moniliform filaments, — *Sphaerococcus.*

Fronds compressed, pinnately branched, ultimate ramuli opposite, alternately short and long, central hyphae of large chains of cells, ramuli cortical throughout, — *Ptilota.*

Fronds cartilaginous, pinnately branched, branches of irregular length, ultimate ramuli opposite, monosiphonous, uncorticated, — *Plumaria.*

35. Base discoid, stipitate branches leafy, with midrib and veins, often proliferous, parenchymatous with prosenchymatous veins, — *Delesseria* (in part).

Base discoid, stipitate branches leafy, deeply sinuous, with midrib and veins, — *Phycodrys.*

36. Base discoid, stipitate, branches ribbon-like with midrib and veins, often proliferous, parenchymatous with prosenchymatous veins, — *Delesseria* (in part).

Base discoid, sub-stipitate, alternate pinnate, dentate branches, cartilaginous, parenchymatous, — *Odonthalia.*

37. Base discoid, stipitate, cartilaginous, palmately branched, central hyphae, cortical parenchyma, superficial moniliform filaments, cystocarps embedded, — *Chondrus.*

Base discoid, stipitate, cartilaginous wedge-shaped branches, central hyphae and superficial moniliform filaments, cystocarps in superficial warts,   *Gigartina.*

Base discoid, stipe short, obovate or reniform, central hyphae and cortical parenchyma becoming short-celled outwards,   *Callymenia.*

Base discoid, stipe branched, wedge-shaped, fronds proliferous, cartilaginous, cortical parenchyma becoming small-celled outwards,   *Phyllophora.*

KEY TO THE COMMON GENERA OF THE HIGHER BASIDIOMYCETES

1. Hymenium exposed from the first, spread over a smooth surface, pores, teeth or woody gills. Fruiting body rarely fleshy (Aphyllophorales)   2.

   Hymenium at first enclosed, becoming exposed over the surface of fleshy gills or pores (Agaricales)   17.

   Hymenium enclosed within a peridium at maturity (Gasteromycetales),   49.

2. Hymenium composed of tubes,   3.

   Hymenium not composed of tubes,   9.

3. Tubes forming a distinct layer,   4.

   Tubes not forming a distinct layer but homogeneous with the pileus,   7.

   Tubes separate from one another, pileus fleshy, juicy,   *Fistulina.*

4. Fruiting body resupinate,   *Poria.*

   Fruiting body, stalked or sessile,   5.

5. Fruiting body fleshy or leathery,   *Polyporus.*

   Fruiting body woody or corky,   6.

6. Surface of fruiting body having a rigid shining crust,   *Ganoderma.*

   Surface of fruiting body hard, woolly or velvety   *Fomes.*

7. Tubes deep and well developed,   8.

   Tubes very shallow or becoming toothed,   *Irpex.*

8. Tubes forming from the centre outwards,   *Polystictus,*

   Tubes regular, rounded or oblong,   *Trametes.*

   Tubes irregular, sinuous or labyrinthiform   *Daedalea.*

9. Hymenium spread over veins, anastomosing
   pores or quite smooth,   10.

   Hymenium spread over the surface of protu-
   berances,   12.

   Hymenium spread over a ribbed surface or
   smooth,   15.

10. Fruiting body sessile, or resupinate, gelatinous,
    with anastomosing veins,   *Merulius.*

    Fruiting body resupinate or erect, waxy,   11.

11. Spores white,   *Phlebia.*
    Spores coloured,   *Coniophora.*

12. Fruiting body stalked or sessile, fleshy or corky, *Hydnum.*
    Fruiting body resupinate,   13.

13. Flesh and spores coloured,   *Hypochnus.*
    Flesh not coloured, spores white,   14.

14. Spines obtuse, deformed and irregularly scat-
    tered,   *Radulum.*

    Spines conical,   *Odontia.*

    Spines very minute,   *Kneiffia.*

15. Fruiting body composed of flat plate-like
    branches,   *Sparassis.*

    Fruiting body stalked or sessile,   *Stereum.*

    Fruiting body incrusting, irregularly branched, *Thelephora.*

    Fruiting body resupinate,   16.

16. Cystidia present,   *Peniophora.*
    Cystidia absent,   *Corticium.*

17. Hymenium lining pores,   18.
    Hymenium covering the surface of gills,   20.

18. Pileus covered with imbricate scales,   *Strobilomyces.*
    Pileus not covered with scales,   19.

19. Tubes long,   *Boletus.*
    Tubes short,   *Boletinus.*

20. Fruiting body fleshy, 21.

Fruiting body rapidly putrefying or shrivelling up, thin and sometimes transparent, 44.

Fruiting body tough or woody, 45.

21. Fruiting body with ring and/or volva present, 22.

Fruiting body with ring and volva absent, 28.

Fruiting body with arachnoid or filamentous veil, 43.

22. Ring and volva both present, 23.

Ring present, no volva, 24.

Volva present, no ring, 27.

23. Spores white, *Amanita.*

Spores black *Anellaria.*

24. Gills free, 25.

Gills adnexed or slightly decurrent, 26.

25. Spores white, *Lepiota.*

Spores pink, *Annularia.*

Spores purple, *Psalliota.*

26. Spores white, *Armillaria.*

Spores brown, *Pholiota.*

Spores purple, *Stropharia.*

27. Spores white, *Amanitopsis.*

Spores pink, *Volvaria.*

Spores brown, *Acetabularia.*

Spores purple, *Chitonia.*

28. Gills mucilaginous or white waxy, 29.

Gills exuding milky juice when broken, spores white, *Lactarius*

Gills neither waxy nor exuding milk, 30.

29. Spores white, *Hygrophorus.*

Spores black, *Gomphidius.*

30. Stem excentric or absent,   31.
    Stem centric, present,   32.

31. Spores white,   *Pleurotus.*
    Spores pink,   *Claudopus.*
    Spores brown,   *Crepidotus.*

32. Stem fleshy,   33.
    Stem cartilagineous or stringy,   39.

33. Gills free,   34.
    Gills sinuate,   35.
    Gills decurrent,   36.
    Gills adnate,   38.

34. Spores white,   *Hiatula.*
    Spores pink,   *Pluteus.*
    Spores brown,   *Pluteolus.*
    Spores purple,   *Pilosace.*

35. Spores white,   *Tricholoma.*
    Spores pink,   *Entoloma.*
    Spores brown,   *Hebeloma.*
    Spores purple,   *Hypholoma.*

36. Gills thick, fold-like, anastomosing,   *Cantharellus.*
    Gills thick, separating from hymenophore,   *Paxillus.*
    Gills thin,   37.

37. Spores white,   *Clitocybe.*
    Spores pink,   *Clitopilus.*
    Spores brown,   *Flammula.*

38. Fruiting body rigid, brittle, generally brightly
        coloured,   *Russula.*
    Fruiting body parasitic on other Agarics,   *Nyctalis.*

39. Gills free or adnate,   40.
    Gills sinuate,   41.
    Gills decurrent,   42.

40. Spores white,   *Collybia.*
    Spores pink,   *Leptonia.*
    Spores brown,   *Naucoria.*
    Spores purple,   *Psilocybe.*
    Spores black,   *Panaeolus.*
41. Spores white,   *Mycena.*
    Spores pink,   *Nolanea.*
    Spores brown,   *Galera.*
    Spores purple,   *Psathyra.*
    Spores black,   *Psathyrella.*
42. Spores white,   *Omphalia.*
    Spores pink,   *Eccilia.*
    Spores brown,   *Tubaria.*

43. Veil forming fugacious arachnoid ring on the
    stem. Spores brown,   *Cortinarius.*
    Veil continuous with the surface of the pileus.
    Spores brown,   *Inocybe.*

44. Gills autodigesting, spores black,   *Coprinus.*
    Gills not autodigesting, spores brown,   *Bolbitius.*

45. Stem excentric or absent,   46.
    Stem centric,   47.

46. Gills simple, dry fruiting bodies reviving in
    water. Spores white,   *Marasmius.*
    Gills branched, dry plants not reviving in
    water. Spores white,   *Xerotus.*

47. Gills anastomosing,   *Lenzites.*
    Gills simple,   48.

48. Edge of gill entire,   *Panus.*
    Edge of gill toothed,   *Lentinus.*
    Edge of gill split longitudinally,   *Schizophyllum.*
    Edge of gill blunt and rounded,   *Trogia.*

49. Peridium globose,   50.
    Peridium bell-shaped, enclosing peridida   56.

50. Peridium composed of a single layer,   *Scleroderma.*
    Peridium composed of two or more layers,   51.

51. Fruiting body gelatinous, rupturing to expose
      stalked receptacle,   52.
    Fruiting body fleshy, opening by an apical pore,   53.

52. Gleba attached throughout to the receptacle,   *Mutinus.*
    Gleba only attached to the receptacle at the apex,   *Phallus.*

53. Exoperidium composed of a single layer,   54.
    Exoperidium composed of two layers, one split-
      ting into star-shaped segments,   *Geaster.*

54. Capillitium threads attached to the peridium,   *Lycoperdon.*
    Capillitium threads free,   55.

55. Sterile base present,   *Bovistella.*
    Sterile base absent,   *Bovista.*

56. Peridiola without funiculi,   *Nidularia.*
    Peridiola with funiculi,   57.

57. Peridium composed of two layers,   *Crucibulum.*
    Peridium composed of three layers,   *Cyathus.*

INDEX